THE
KILLING
ZONE

THE
KILLING
ZONE

THE GRENADIER GUARDS
PUSHED TO THE LIMIT ON
HELMAND'S FRONT LINE

Lt Col RICHARD DORNEY

EBURY
PRESS

1 3 5 7 9 10 8 6 4 2

First published in 2012 by Ebury Press, an imprint of Ebury Publishing
A Random House Group company

The Random House Group Limited Reg. No. 954009

Addresses for companies within the Random House Group can be found at
www.randomhouse.co.uk

A CIP catalogue record for this book is available from the British Library

The information in this book is believed to be correct as at 1 May 2012,
but is not to be relied on in law and is subject to change. The author and
publishers disclaim, as far as the law allows, any liability arising directly or
indirectly from the use, or misuse, of any information contained in this book.

The Random House Group Limited supports the Forest Stewardship
Council® (FSC®), the leading international forest certification organisation.
All our titles that are printed on Greenpeace approved FSC® certified paper
carry the FSC® logo. Our paper procurement policy can be
found at www.randomhouse.co.uk/environment.

Designed and set by seagulls.net

Printed and bound in Great Britain by CPI Group (UK) Ltd, Croydon, CR0 4YY

ISBN 9780091948863

To buy books by your favourite authors and register for offers visit
www.randomhouse.co.uk

DEDICATION

This book is dedicated to those men of the 1st Battalion Grenadier Guards who gave their lives in Afghanistan:

Guardsman Simon Davison 3 May 2007

Guardsman Daniel Probyn 26 May 2007

Guardsman Neil 'Tony' Downes 9 June 2007

Guardsman Daryl Hickey 12 July 2007

Guardsman David Atherton 26 July 2007

Guardsman Jamie Janes 5 October 2009

Warrant Officer Class 1 (RSM) Darren Chant 3 November 2009

Sergeant Matthew Telford 3 November 2009

Guardsman James Major 3 November 2009

Lance Sergeant David Greenhalgh 13 February 2010

CONTENTS

HELMAND PROVINCE

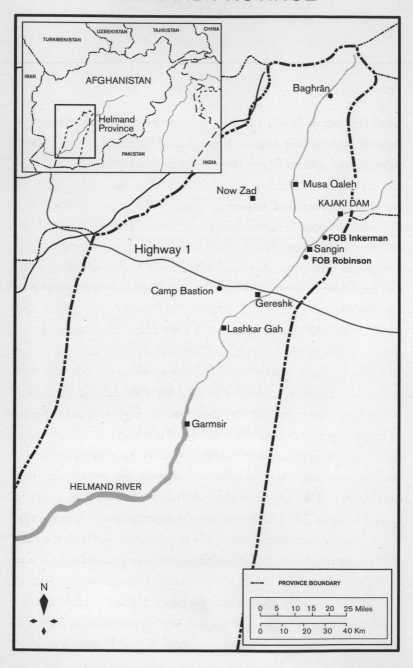

INTRODUCTION

In the spring of 2007, the situation in southern Afghanistan was very different to the state of affairs found there today. Since then, tremendous advances have been made in security, stabilisation and development. At the time of writing, the Afghan government has started the process of assuming control of its own security from the International Security Assistance Force (ISAF). It is perhaps worth remembering that in 2006 little more than a British battle-group was sent to tie down a huge area of Helmand. More than five years later a formation the size of a division is still fighting to control the same area. The addition of 20,000 US soldiers and Marines in 2009 has made a huge difference. The British force, now in excess of 10,000 troops, is concentrated in central Helmand and has the force density levels to bring about the changes that were planned so long ago. In 2007, 12 Mechanised Brigade helped to lay the foundations that have allowed the Afghan security forces to regain control in much of southern Helmand.

Many thousands more Afghan soldiers have been recruited and trained, and they are better equipped and supported than ever before. The Afghan National Police, virtually non-existent in Helmand in 2007, are now present in ever-growing numbers. While these government forces are by no means perfect, they are vastly improved and will be charged with responsibility for the future security of the country. Cultural differences, war weariness and Taliban activity ensure that mentoring these forces remains a significant challenge, although the improvement in their capability is tangible. The rag-tag band of poorly equipped men

1

encountered by the Grenadier Guards in 2007 has largely been consigned to history.

It is not my intention in this book to question the success or otherwise of the military campaign in Helmand. Nor is there any strategic analysis or political comment; there will no doubt be sufficient volumes of critical comment added to the bookshelves of history by soldiers and academics both. However history records our nation's involvement in Afghanistan, whatever terrible mistakes or astonishing successes are noted, it must be remembered that these were made with the blood of young men and women. Military history always records the names of generals and politicians, but the soldiers who make and break their reputations are rarely adequately documented. Some of the men mentioned in these pages returned to Helmand in 2009 and were subsequently killed or wounded. The purpose of this book is to record their heroism. Men and women from a variety of regiments and corps stood shoulder to shoulder with the Grenadiers and they are an important part of this story too. Their expertise and commitment were invaluable. It is impossible to document every action or significant event of the tour so these pages, in reality, only provide a flavour of what took place. Acts of courage are sometimes rewarded with medals, but more often they go unrecorded. There are never enough medals to go around and the British system of awards is famously conservative.

When writing this book I was reminded of a conversation I had with a young soldier in Helmand. He had narrowly escaped death in an IED strike in which his friend had been killed and was due to return to Sangin and the scene of the attack. 'I don't want to go back there, sir, I am shit-scared,' he said. After a pause he added, 'But there's no way I am not going, because my mates are there and there's no way I would jack them.' To me this spoke volumes about the fighting spirit and courage of the British soldier. He did go back and served with distinction in many other encounters with the enemy, just as other generations of Grenadiers have done in our nation's history. In a few years, the word 'Helmand' will probably disappear from our newspapers and national consciousness, but it will live on in other ways. It will not be forgotten by the families of those who gave their lives there. It will remain in the thoughts of those who bear the physical scars of war and by those who are haunted by their memories. While the book is dedicated to those who gave their lives in Afghanistan, we should continue to support the living who will be affected by the consequences of war for years to come. The publishers will donate the author's profit from this book to the Colonel's Fund Grenadier Guards and to other service charities in support of this aim.

Richard Dorney
Surrey, 2011

1

ALL IN A DAY'S WORK

6 Platoon had been clearing the compounds and buildings in the village of China to the south of Kajaki in Helmand Province for several hours. They had left their forward operating base in the early hours of the morning and had so far seen no sign of the Taliban in the deserted and dusty hamlets. Dawn had broken and the first rays of the Afghan sun were now illuminating the spectacular landscape. The mountains of the Hindu Kush dominated the horizon and contrasted with the green poppy fields which now stretched out in front of 6 Platoon's Grenadiers. A prominent wadi or dry river bed marked the line of departure for the next phase of the operation and the lead section from the Afghan National Army (ANA), now sat in the bottom of it, smoking. Among them, their British mentors from the Grenadier Guards peered nervously through the undergrowth at the ground ahead. It was 21 April 2007 and the Grenadiers had been in Kajaki for only a couple of weeks. The previous day, 6 Platoon had been ambushed in another village nearby so they were understandably vigilant today. A British platoon from the 1st Battalion Royal Anglian Regiment was moving to the right of the Grenadier and ANA platoon. The lead ANA section now rose and moved out of cover and into open ground in the direction of the red-topped poppy fields. They were joined by Captain Alexander Allan, who was temporarily attached to 6 Platoon. Second Lieutenant Howard Cordle, the platoon commander, was now in the wadi

with the second Afghan section and Sergeant Ty-Lee Bearder, the platoon sergeant, was further back coordinating the move of the Afghan platoon. To the rear left of the platoon, more Anglians were occupying a piece of high ground overlooking the area now being crossed by the other two platoons.

Unbeknown to the forward troops, the British soldiers on the high ground had spotted movement in the poppy fields to the front of 6 Platoon. Further observation confirmed that the movement was in fact several Taliban fighters who appeared to be withdrawing from the British with some haste. As a warning was being passed over the radio suddenly all hell broke loose, and the early morning silence was shattered by the loud cracking sound of high velocity bullets as they sprayed across the open ground and through the undergrowth. The ground around Allan's feet was churned by the bullets fired from Russian-made AK-47 rifles. Allan and the ten or so other members of the exposed section threw themselves to the ground and tried to make themselves as small as possible. As they did so, green and red tracer rounds spat through the air in great numbers adding to the shock effect. The tracers were coming in from several different angles and criss-crossed in the air over their isolated targets. The fire was so intense that there was no possibility of any of those now trapped in the open ground being able to run or even crawl into cover. They could go neither forwards nor backwards and Allan could be seen curled into a foetal position on the dusty ground. Small dust clouds engulfed him as fresh bullets struck all around him. Cordle was in a safer position in the wadi, but even here the Taliban bullets were ripping through the trees and showering those below with leaves and wood splinters.

Cordle tried to take stock of what was happening. There was a great deal of confusion and the intensity of fire indicated that this was a well-prepared ambush by an enemy with strength. The ANA and Grenadiers in the wadi now returned fire towards the origins of the bright tracers which continued to crack overhead at an incredible rate. Sergeant Bearder had moved his rear section of Afghans into a shallow ditch from which he peered at the scene ahead. Bullets shredded the undergrowth around him and small branches from above his head were severed by bursts of machine gun fire. He tried to talk to his platoon commander on the radio but this was difficult given the deafening noise around him; his Afghan troops were now firing long bursts into the distance. Bearder could plainly see the trapped Captain Allan and his Afghan soldiers, dust clouds flicking up around them. Something had to be done quickly or none of the lead section would survive. He looked around to try and find an answer to this deadly dilemma but there was no simple solution, the enemy had planned well. As he looked to his right he saw a small piece of raised ground, about 150 metres away. Tracer rounds were still cutting the air between the two positions. The piece of ground was probably no more than a metre higher than his current position but it would allow much better observation of the enemy positions. Bearder decided that from the small bank he would be able to suppress the enemy and perhaps give the surviving members of the lead section a chance to withdraw from the killing zone. He quickly passed the plan to 'T', the platoon sniper who was with him, and to Cordle who was directing the ANA from the wadi while simultaneously attempting to call for air support. Bearder and the sniper broke cover, leaving the Afghan section behind.

They ran as hard and as fast as they could but hardly seemed to be moving due to the heavy weight of their body armour. Tracer rounds whipped through the air around the two sprinting men and the intensity of fire increased as the enemy fighters spotted them and switched their fire to these two exposed targets. After what seemed like hours but was in reality merely minutes, the two reached the low bank, which now looked even smaller.

Bearder wondered if he had made the right decision as it was now apparent that this small rise provided only a very slightly improved view of the enemy positions.

By now most of the Taliban to the front had identified Bearder's position and were firing at the exposed bank. Tracer rounds continued to crack overhead and to the sides as other bullets thudded into the bank, throwing up clods of earth and sand. Looking through the scope on his L96 sniper rifle, T could see loopholes that had been cut into the earth walls of an adjacent compound. Puffs of smoke regularly came from the loopholes signifying that these were well-prepared Taliban ambush positions. He took careful aim and fired into the holes, concentrating hard to aim the shots into the small slots. He repeatedly worked the bolt of the rifle and continued to pump round after round at the attackers. Bearder quickly loaded a 40mm high explosive grenade into the launcher fitted underneath his rifle. He took aim and fired at the compound, there was a sharp 'pop' and a slight delay before a black cloud signalled a hit on the target. The enemy fire temporarily slowed and the two Grenadiers continued to fire at the concealed positions.

The temporary lull in enemy fire as the Taliban switched onto Bearder and T was enough to allow the exposed lead section to withdraw the 60 metres or so back into the wadi. The relieved ANA soldiers threw themselves into the sandy river bed and lay there panting; they were quickly joined by a shaken Allan who rapidly found Cordle. Allan and the Afghans were still alive because Bearder and T had bought the few seconds of respite needed by distracting the enemy. But they were now under intense fire as every enemy position engaged the only visible targets on the top of the little bank. Cordle was quick to radio to request air support. He called in the coordinates of the enemy positions and described the target to the pilot. When the young officer looked up from his map and turned towards the little bank, he was horrified that

Bearder and T could no longer be seen on top of the mound. The sniper's rifle was on its bipod and the whole of the bank was being raked by machine gun fire. The two Grenadiers were in fact lying flat on the ground behind the bank. An enemy PK machine gun had opened up and the rate of fire was so intense that the two men had been forced to crawl backwards into cover. The ANA and remaining Grenadiers poured fire in the direction of the Taliban gunner and, as he watched, Cordle saw two crawling figures move back onto the bank where they commenced firing again.

This was the pattern for the following hour or so until coalition aircraft appeared overhead. Several 250lb bombs were dropped onto the occupied compounds to the front and huge, angry-looking brown and black clouds burst out of the buildings, expanding in all directions until the whole area was shrouded in brown dust. The enemy had either crept away or had been decimated by the aerial bombs. The contact had lasted for several hours and as silence once again descended on the green fields, Bearder and T made their way back to the exhausted ANA platoon. Bearder's first concern was to establish how many casualties the platoon had taken and to ensure that the wounded were treated. He was relieved to find that none of the Grenadiers was injured, but was astounded that there were no wounded Afghans either. It had been a miracle that the exposed lead section had returned to the wadi without even a scratch. 6 Platoon were now ordered to withdraw along with the Royal Anglian troops who had been in contact to their right. As they passed through the Anglian company sergeant major's position, Bearder was very surprised to hear that there too not a single UK or Afghan soldier had been hit during the intense gunfight. There could be no doubt that many enemy fighters had died in the air strike, but there was no sympathy for them.

The exhausted troops made their way back to Kajaki where they were able to drop their heavy equipment and strip off their sweat-soaked T-shirts. As the Grenadiers sat around refilling magazines and checking equipment, there was a buzz of excitement in the air. It had been an incredible experience, intensely frightening but at the same time exhilarating. The atmosphere in the relatively safe environment of the forward operating base was jovial with much banter and bravado. Privately, however, everyone knew they had been incredibly lucky. The Grenadiers were only a couple of weeks into their operational tour of Helmand Province and they had a full six and a half months in front of them. If this was going to be the routine, they would be lucky to survive. They would have to learn some tough lessons, and fast, if they were to match the deadly cunning of their enemy. Some of these Taliban fighters were probably walking across the mountains right now, in the direction of Kajaki, where they would lay more ambushes for the British and Afghan troops.

2

IN THE BEGINNING

The events of 11 September 2001 changed the world for ever. For those of us who watched the unfolding tragedy live on TV it was obvious that war would be the inevitable result. The US was sure to retaliate and would use all of its awesome military power to punish those who had murdered so many on that dreadful day. It was not immediately clear who had been responsible for the worst terrorist atrocity of modern times, but the finger of suspicion was already pointing towards the dusty interior of Afghanistan. This south central Asian country had long been politically unstable and under the Taliban government Afghanistan had become a terrorist haven where training camps had been operating for some time. The Taliban had emerged in the south of the country around 1994. It was a murky organisation largely comprised of religious students recruited from the madrassas of Pakistan. With the backing of Pakistani elements, the movement was able to gain momentum, seizing control of large swathes of the countryside, until by 1996 Kabul and around 90 per cent of the country was under their influence. Between 1996 and 2001, the Taliban allowed extremist Islamic terrorist groups to establish themselves inside the country's borders. From here they exported hatred and extremism around the globe. Under the protection of the Taliban, Osama Bin Laden fermented the extremist ideal that became al-Qaeda. This ruthless and fanatical group was to be responsible for the murder of thousands of innocents around the world.

In the weeks that followed 9/11, the US intelligence machine went into overdrive and the military prepared for war. It didn't take long for the Americans to discover the source of the callous attacks against them sheltering inside Afghanistan. For those of us serving in the British Army at the time, we wondered what the future held. Many Britons had been killed in the World Trade Center on 9/11 and it seemed likely that the UK government would support the US just as it had done over the invasion of Kuwait by Iraq in 1991. An ultimatum was issued to the Taliban; they could give up Bin Laden and his lieutenants or face the consequences. The Taliban declined the opportunity to cooperate and subsequently faced the full force of the US and coalition military.

The opening attacks against al-Qaeda and the Taliban came in October 2001 under the auspices of the US-named Operation Enduring Freedom. These were the first blows in what was to become a protracted and costly campaign. The UK was involved from the very beginning and, as predicted, threw its entire support behind its US ally. The British named the initiative Operation Veritas, a characteristically uninspiring title for what was to be a very considerable military commitment. The first British troops on the ground joined their US allies in late 2001. In the weeks that followed, our TV screens were filled by images of hi-tech war machines delivering death and destruction to an almost medieval enemy. Bin Laden and his al-Qaeda followers hid in the Tora Bora caves of the White Mountains of eastern Afghanistan and were relentlessly attacked. This was a truly asymmetric war; the world's greatest superpower and its allies versus an indeterminate number of fanatical religious terrorists.

The Taliban government was quickly overthrown by the Northern Alliance, an anti-Taliban faction, which was supported by professional western troops. The surviving al-Qaeda terrorists evaporated into the countryside, many walking through the myriad of mountain passes into the tribal areas of neighbouring Pakistan. With the Taliban removed and al-Qaeda dispersed, Afghanistan was now a political blank canvas onto which the West could paint a democratic template for stability. The country had known nothing but war for more than 20 years. The Soviet Union had invaded in 1979 and for almost a decade a bloody and ruthless campaign had been waged between the Soviet invaders and the Mujahedeen resistance fighters. When the Soviets withdrew a decade later, a disparate collection of political and tribal groupings headed by a succession of warlords and religious leaders all vied for control of the country. The Taliban eventually established themselves in 1996. Their rule was ruthless, barbaric and medieval in character, with public executions becoming commonplace. The people of Afghanistan had been brutalised and had lived in abject poverty, intense fear and repression for much of the population's living memory. There was now a real opportunity to improve life for the citizens of one of the poorest countries on earth and to stabilise a volatile region.

It was essential not to allow the Taliban to regroup or to allow extremist tyranny to return. Reconstruction was the key to stability. New roads, electricity, water and schools would convince the population that peace was possible. If the infrastructure could be improved, the economy would follow and with it normality could be brought to the troubled country. In December 2001, the International Security Assistance Force (ISAF) was authorised by a United Nations Resolution. This multinational military force was deployed to Kabul and the surrounding area with the aim of providing a safe and secure environment for reconstruction. These efforts, although not without opposition, were largely successful and some local stability was established. In 2003, NATO took over command of ISAF with UN authority. Now with improved resources it was possible to expand security and reconstruction efforts into other parts of the country. Unfortunately, events in

Iraq, where another war had begun, were to eclipse achievements for the next three years. British and coalition forces continued their efforts in Afghanistan, but our TV screens were filled with the images of another unfolding tragedy in the Middle East.

Between 2003 and 2005, ISAF executed a staged expansion programme into the countryside, and particularly in the west of the country. The reconstruction effort and the development of the economy were initially slow, largely due to the complete lack of infrastructure within the country. The presence of international forces did allow political progress, including extending the writ of the Afghan government into the provinces by providing security and reassurance. Democratic elections were successfully held in 2004 and parliamentary elections followed in 2005. The UK provided over 1,000 troops to ISAF at this time. However, the Afghan government had little influence in the south and east of the country and in 2006 plans were drawn up to extend the reconstruction efforts into these volatile and remote areas. The southern and eastern provinces of Afghanistan were traditionally lawless and difficult to govern. Support for the Taliban was stronger in the south and the proximity of the Pakistan border was a further complication. This area was also at the heart of the narcotics trade. Hundreds of tons of opium extracted from the poppy fields of Helmand were transported through Pakistan and onto the streets of Europe every year. There was an historic mistrust of outsiders in the south and there were many with a vested interest in opposing government rule. International security forces now sought to stabilise the provinces of Helmand, Kandahar, Uruzgan and Zabol. This would be done by establishing Provincial Reconstruction Teams as they had done successfully in other provinces. Some 16 nations contributed troops with the US, UK, Canada and the Netherlands taking the lead. The main British focus would be in the province of Helmand, where Taliban influence and the rugged desert terrain would make reconstruction a difficult task to achieve. In January 2006, the then Secretary of State for Defence, John Reid MP, gave a speech in which he now famously stated that British troops were there to help Afghan reconstruction and 'would be perfectly happy to leave without firing a shot'. Some five years later, the British death toll was approaching 400. Hundreds more had been wounded in southern Afghanistan.

Helmand Province lies to the south of the formidable Hindu Kush mountains. It is bordered in the north by Ghowr Province and by Nimroz and Farah to the west. Kandahar, Uruzgan and Daikondi mark its eastern geographical limits. The northern part of Helmand rises rapidly into a rugged mountain range with steep valleys and passes, many of which are completely inaccessible to motor vehicles. The rugged Chagai hills to the south sit astride the 160km-long border that separates Helmand from the Pakistani province of Baluchistan. The Helmand River, Afghanistan's longest, flows from the high ground in the north, following a southerly route through the otherwise barren desert lowlands before swinging away to the south-west. This area of Afghanistan has suffered drought throughout its troubled history, which means that civilisation there has grown up dependent on the irrigation ditches that draw their lifeblood from the Helmand River which flows all year round. These cultivated areas are often home to lush vegetation, in stark contrast to the barren desert that sandwiches the river. For obvious reasons this became known to the British Army as the Green Zone. These relatively fertile areas provide an ideal home for the notorious Helmand poppy, source of so much heroin-induced misery in western Europe and the rest of the world.

The provincial capital is the busy town of Lashkar Gah, close to the banks of the Helmand River and just south of the main arterial highway linking Farah and Herat in the west with Kandahar and Kabul in the east. This vital tarmac road, known simply as Route 1, has been key to the redevelopment of southern

Afghanistan. Once you turn off this major route, you are hard pressed to find many other metalled roads.

So this is Helmand: barren, short of water, difficult to transit and in the summer as hot as hell. If this sounds inhospitable, it may be worth considering that it is also one of the most seismically active areas of the world and is prone to earthquakes. It is littered with land mines, sown by an increasingly desperate and ill-disciplined Soviet Army as it retreated. Few of these minefields are recorded and they have claimed hundreds of lives over the last 20 years. Add to this the fact that neighbouring Kandahar is regarded by many as the spiritual home of the Taliban, and it is easy to see why many people were concerned about this latest phase of the ISAF expansion programme.

The British Royal Engineers were sent into Helmand first to establish a base from which Task Force Helmand could operate. The main force, which followed later, was based upon elements of the HG of 16 Air Assault Brigade and an airborne infantry battlegroup. These troops pushed out into the desert towns of the province in order to lay the foundations for reconstruction. They were to be joined by representatives from the Department for International Development, the Foreign and Commonwealth Office and non-governmental organisations. These agencies would decide how best to deliver aid to a needy population and would prioritise reconstruction projects.

From the very beginning, Task Force Helmand met determined resistance from the Taliban and from those who fell under their influence. The soldiers from 16 Air Assault Brigade repeatedly found themselves isolated and faced frequent attacks. Resources were stretched and British casualties mounted. The UK public was slow to notice what was happening in southern Afghanistan, largely because there were few journalists there at the time. As the death toll rose, Afghanistan once again returned to our TV screens and newspapers. Desperate accounts of soldiers fighting with limited resources galvanised public interest. The summer of 2006 was marked by very high casualty figures and the realisation that one infantry battlegroup would not be sufficient for the task in Helmand. It was clear that the Taliban would not relinquish their heartland without a very determined fight. Likewise, those involved in the trafficking of opium also had a great deal to lose by letting the British into their backyards. At the same time, the Canadians too were experiencing fierce resistance in Kandahar Province, especially along the Helmand border. It was now becoming apparent that a major rethink would be required if ISAF were to bring stability to the south.

As the summer of 2006 passed into autumn the weary paratroops handed over their battlegrounds to the Royal Marines of 3 Commando Brigade. The Marines were substantially reinforced and the Afghanistan operation, known as Operation Herrick, had become a full brigade-sized operation. The arrival of the winter snow in the highlands, which hinders Taliban movement, and the additional troops available meant that the Commandos were able to build upon the costly foundations laid by 16 Brigade. There were 39 UK fatalities in Afghanistan in 2006 and things would get worse in 2007. Many valuable lessons had been learned and more efforts were now directed at training the Afghan Security Forces. Intelligent targeting by the Marines meant that some key Taliban commanders were removed from the scene. The Taliban had been given a bloody nose by the Paras and the Marines used the winter to consolidate gains and prevent enemy preparations for the expected spring offensive.

The events taking place in Afghanistan during the winter of 2006/7 were followed keenly in the UK by the soldiers of 12 Mechanised Brigade. 12 Brigade were to relieve the Marines in the spring and it would be they who would have to face any Taliban resurgence as the weather improved. For most of the brigade units, this would be their first time on operations for almost two years. This was not the case for the soldiers of the 1st Battalion Grenadier Guards; they had followed the events of the summer of 2006 from the Iraqi desert.

The Grenadiers had long been scheduled to join 12 Mechanised Brigade for its Afghanistan tour, but had been sent to Iraq at short notice. The knock-on effect was that the Grenadiers would have to turn themselves around in only six months. This would be a very tall order and would leave little time for training, not to mention the strain that it would place upon the families and the battalion's administrative structures. This quick turn-around was exceptional and those responsible for sending the Grenadiers to Iraq at such short notice must have been only too aware of the difficulties that this would cause. As the Marines stepped up their operations in Helmand in an attempt to prevent a spring offensive, the Grenadiers had to knock themselves into shape for a new mission. Planning for the Afghanistan tour took place despite the fact that the battalion was simultaneously facing an equally deadly enemy in the sands of southern Iraq.

Preparation for an operational tour is always a frenzied time, full of activity and uncertainty about the task ahead. This tour was to be even more difficult because the Grenadiers had been designated as the Operational Mentoring and Liaison Team (OMLT). This meant that they would be responsible for organising and training the ANA in Helmand, as well as deploying on operations with them. The Grenadiers would work alongside their British colleagues in the rest of 12 Brigade, but they would have to ensure that their ANA charges were capable of conducting joint Afghan and UK operations. No one had really done anything similar on this scale before and the road ahead would have to be slowly found by trial and error.

Organising the battalion for the allocated mission was a difficult and confusing task. What skill sets would be required? What of the rank structure? What equipment would be required to fulfil the role? There were a thousand questions to be answered and the only useful guidance available had to come from the Royal Marines in Afghanistan, who were feeling the way ahead and learning daily lessons themselves. It was soon discovered that the 1st Battalion Grenadier Guards OMLT would mentor the 3 (Hero) Brigade of 205 Corps of the Afghan National Army (3/205 ANA). 3 Brigade was organised into five kandaks, the rough equivalent of a UK battalion. Three of these kandaks were made up of infantry soldiers, one was a combat support (CS) unit and the last was a combat service support (CSS) kandak, responsible for medical support and logistics. It was clear that the Grenadiers would have to organise themselves along similar lines in order to be able to mentor the fledgling Afghan brigade. Additional expertise would be required to train the remaining two specialist support kandaks, so troops from other 12 Brigade units were attached to the Grenadiers; gunners from the Royal Artillery, sappers from the Royal Engineers, reconnaissance specialists from the Light Dragoons and men from the Royal Logistic Corps and Royal Electrical and Mechanical Engineers (REME) would form the CS and CSS mentors.

The Afghan brigade headquarters too would need mentors and these would be found from within the Grenadiers' own battlegroup headquarters. The commanding officer, Lieutenant Colonel Carew Hatherley, would be responsible for mentoring the 3/205 brigade commander in addition to commanding his own British troops. The remaining Grenadier HQ staff would 'pick up' their Afghan counterparts. Mentoring would be a rank-heavy task and the officers and NCOs would have to use all of their military experience and patience to train their respective kandaks. The UK OMLT companies would each be about 35 strong, which meant that some of the younger soldiers could not be used for the mentoring mission.

In all, about 200 Grenadiers were dedicated to this task and they would have to learn a whole host of skills. Not least of these was extensive training in the Afghan culture and language. It was important to have an understanding of the Pashtunwali, roughly translated as 'the way of the Pashtun'. The Pashtuns are the predominant ethnic group in Afghanistan and have been the principal political group for some time. Most of Helmand Province was Pashtun, with the southern part being Beluchi. There are a bewildering number of tribes and corresponding languages spoken in Afghanistan, but there are two official languages: Dari and Pashto. At the outset, all of this was puzzling to young men who had only just started to grasp the intricacies of the Arab culture in Iraq.

Those men not required for the OMLT task were detailed for other jobs within the brigade. The commanding officer decided to form an additional rifle company using these men, which would be used for manoeuvre and ground-holding tasks along with the other infantry units. It was unlikely that this company would be under his command but it would provide the brigade with additional troops and flexibility. This meant that the available manpower not used for the OMLT would be formed into a single company. This may sound like a straightforward job to those who have never experienced it, but it was in fact an extremely difficult and at times frustrating task for those involved. The company would be commanded by Major Will Mace of the Scots Guards, something he had not anticipated when he was posted to the Grenadiers only weeks earlier. In spite of their unsatisfactory start in life, members of the new 3 Company quickly gelled and overcame everything that was thrown at them. Because of their different role, the company had to be trained along more conventional lines, which meant a separate programme and timetable for deployment.

It was also decided that the brigade would train an additional reconnaissance force which would be known as 12 Brigade Reconnaissance Force (12 BRF). In October 2006, 160 volunteers formed up in Lille Barracks, Aldershot, where they were to attend a selection course. This course lasted for two weeks and was necessarily arduous in nature. There was a great deal of physical training and many of the essential military skills that would be required in Helmand were tested under very demanding circumstances. The volunteers were whittled down to 85, well over half of whom were Grenadiers, including the company sergeant major and the second in command. Many of the BRF soldiers possessed specialist skills and came from other arms and corps of the brigade. There were sappers, gunners, REME, Light Dragoons and King's Royal Hussars personnel all represented in the company. This disparate but highly motivated organisation was to be commanded by Major Rob Sergeant of the Coldstream Guards. Because of its specialist role, 12 BRF had yet another entirely separate training and deployment programme to that of the OMLT and 3 Company. This was a further complication to administration and a nightmare for the quartermaster who was responsible for equipping and sustaining all of these sub-units on their separate ventures.

The dispersion of the battalion was still not yet complete. Because the 1st Battalion Grenadier Guards was a well-manned and well-recruited unit, they were detailed to reinforce others. A Territorial Army (TA) company was mobilised in order to assist 12 Brigade with operations in Helmand. Somme Company of the London Regiment joined the Grenadiers for training and although they would not be under Grenadier command, they were reinforced with a Grenadier platoon. From the very beginning Major Milan Torbica, the company commander, and Hatherley were determined to make the deployment a success. Torbica was able to place one of the Grenadier sections into each of his TA platoons, which allowed him to form a fourth platoon and to spread the experience around the company. Most of the TA soldiers had also seen service in Iraq and they knew their trade. A strong Grenadier thread now existed within the company and each man, Guardsman or TA soldier, wore the blue and red shoulder flash of the Household Division. Somme Company would soon find itself in the thick of things.

The Grenadiers also provided three infantry sections and a couple of 81mm mortar teams to the 1st Battalion Royal Anglian Regiment, which was to be responsible for operations in the northern part of Helmand. The first time that the various groupings were able to gather collectively was in mid-November when they came together to attend lectures from the Operational Training Advisory Group or 'OPTAG' for short. These presentations were mandatory before every tour but were particularly relevant this time because few present had been to Afghanistan before. The young Guardsmen listened intently to two days of lectures on how the Taliban were currently operating and the lessons that had been learned by fighting them.

After another couple of days with OPTAG, learning or revising some of the practical techniques that would be required, the companies once again went their separate ways. Each of the battalion's sub-units now had to concentrate on its individual mission and how best to achieve it. This meant protracted periods away training in different locations at different times. Much time was spent on the basic skills and drills that would be needed in Afghanistan with a great emphasis on weapons training and first aid. It was clear that the Taliban liked to get in close and would often stand and fight. There were many stories of lengthy gun battles and miraculous escapes from the previous summer. The troops would need to be highly skilled with all of the various weapons in use but in addition they would have to master some less familiar systems. The .50 calibre machine gun and the 40mm grenade machine gun (GMG) were just two examples. In spite of the very obvious dangers that the battalion would face, there was no shortage of volunteers. Many NCOs posted away from the battalion were trying every trick in the book to get back for the tour and the young Guardsmen newly out of training were chomping at the bit to join the Grenadiers for this mission.

At the start of the English winter in November and December, the Grenadiers were on the move once again. This time it was to Norfolk for more OPTAG-led exercises and then to Lydd in Kent, where some intensive shooting training was conducted. Some realistic and useful preparation was done before the battalion returned to Aldershot where they cleaned and accounted for all of their equipment before heading off on a well-earned leave. This was a special time for the families who had missed their loved ones during the summer of 2006 and would have the worry of seeing them go off on operations once again in the early spring. The UK media was now alive to what was happening in Afghanistan and this did nothing to ease the minds of those staying at home who had to endure the graphic TV footage.

Christmas leave was all too short and the first week in January 2007 found the battalion once again preparing for operations. Hatherley and some of his key planning staff had already flown to Afghanistan on a reconnaissance for the mission.

A bewildering array of military vehicles would be used in Afghanistan and drivers had to be trained in their use. The rugged desert terrain and extreme weather conditions meant that driving was a very hazardous occupation. For some this meant lengthy courses away to gain the qualifications necessary to legally drive the heavily armoured vehicles. The man charged with coordinating all of this activity was the battalion second in command or 'senior major' Simon Soskin. Training, courses, deployment and organisation were all his responsibility and given the scattered nature of the battalion and the diversity of tasks, this was no mean feat. Soskin's job was made a little easier by the presence of Regimental Sergeant Major Andrew 'Stumpy' Keeley (so named because of his powerful build). The RSM in any battalion presides over the Sergeants' Mess which provides the essential power that drives the battalion. Updates on current operations and the latest intelligence summaries were studied closely as they would dictate the way in which 12 Brigade would conduct its operations. All of this activity finally came together in a series of brigade-level exercises in January.

As February arrived the brigade moved to Otterburn on the Northumbrian side of the Scottish border for exercise Pashtun Ace. Here the extensive range facilities allowed a much greater freedom of movement and permitted all of the brigade weapon systems to be integrated. Artillery, mortars and even fast jets delivered lethal projectiles onto targets positioned on the windswept hillsides. For most, this was the first time that they had seen live bombs dropped from aircraft but it was definitely not going to be the last.

Although the Grenadiers had endured five months of intensive training, this was far less than any other unit in the brigade, all of which had a much longer lead in time. The condensed time frame had been hard work, but the battalion had achieved everything asked of it. Recent experiences in Iraq had stood the men in good stead during the training. There were similarities in procedures and some in the battlegroup had first-hand recent experience of enemy contact.

The first flights to Afghanistan were scheduled to depart in early March 2007. Some were able to take a short period of leave before reporting for their flights and for the youngest members of the battalion there was a great sense of anticipation at the task ahead. The last time that the battalion would be together until the autumn of 2007 was Friday 9 March. On this day the battalion and many of its families crowded into the Aldershot Garrison Church for a service. The Reverend Stephen Dunwoody, the battalion padre, conducted the service. Everyone present was conscious that it was quite possible, even likely, that there would be fewer members of the battalion sitting in church in seven months' time. Should this be the case, it would be Stephen Dunwoody's unenviable task to despatch the fallen back home for burial.

HELMAND RIVER VALLEY

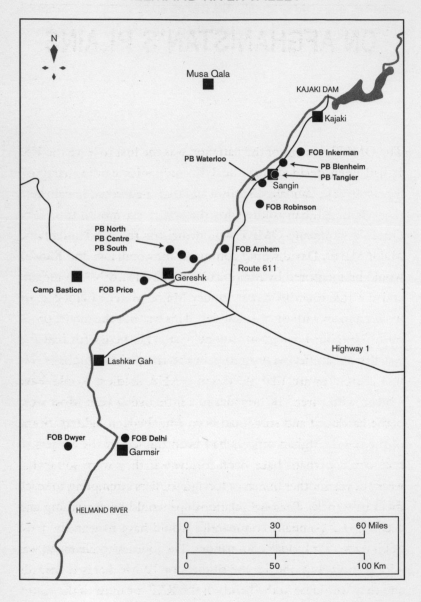

3

ON AFGHANISTAN'S PLAINS

The OMLT element of the battalion was the first to leave the UK. It had been decided that the UK companies would marry off against specific kandaks on arrival and that they would remain with those same Afghan soldiers for the whole six-month tour. The Queen's Company OMLT would mentor the 1st Kandak and Major Martin David would command the company; 2nd Kandak would be mentored by Major Toby Barnes-Taylor's 2 Company; and the Inkerman Company under Major Marcus Elliot-Square would remain with the 3rd Kandak. This was not the model previously used, but Lieutenant Colonel CarewHatherley felt that this was the most effective way to gain the trust of the Afghans. He was acutely aware that these same ANA soldiers would have worked with three UK brigades in a little over a year. Most were battle hardened and some had been Mujahedeen fighters. Some of the senior Afghan officers had been trained by the Soviets so they could perhaps have been forgiven if they were somewhat weary of yet another bunch of foreign soldiers attempting to teach them new tricks. Personal relationships would be everything and each OMLT company commander would have to gain the trust of his respective kandak commander. The journey to Helmand was a long and tiring one for the soldiers of 12 Mechanised Brigade and they would be in the hands of the RAF for most of the route. The process of moving large bodies of soldiers between destinations is a necessarily slow one and is punctuated by roll calls and briefings at various stages. For those who have endured this process many times before it is both tedious and at times infuriating. Stories of long delays and RAF inefficiency usually circulated among the troops and were often embellished each time they were told, sometimes unfairly for the RAF which was forced to maintain a very busy air bridge between both Iraq and Afghanistan. The fleet of aircraft with which the RAF was expected to perform this task was both ageing and small. When one considers the number of troops and essential supplies shipped to theatre, it is easy to see the scale of the task they had been set. In addition to delivering troops they also have to extract those unlucky enough to have become casualties, a vital task that often means the difference between life and death. These facts got little consideration from those who indulged in slating the 'Crabs' and the cultural differences between green and light blue often generated feelings approaching something close to hatred.

The troops were pessimistic and often resigned to long delays even before they left barracks. In fairness to the RAF, great strides have been made in improving the service but no self-respecting soldier was likely to admit to that. The OMLT element of the Grenadier Guards, along with other troops from 12 Mechanised Brigade, arrived at the Air Mounting Centre in mid-March. After the usual checks of passports and documentation, together with the obligatory wait, the troops were eventually loaded onto coaches to the airfield of departure at RAF Brize Norton in Oxfordshire. After another inevitable wait, the weary passengers were finally herded onto their waiting RAF Lockheed L-1011 Tristar transport aircraft. The RAF acquired these ex-commercial passenger planes in the mid 1980s and after so many years they looked very tired. They were none-the-less adequate for the task and the OMLT personnel arrived in Kandahar, southern Afghanistan, safely after a seven-hour flight. The final part of the journey was always unnerving because the passengers were required to don their helmets and body armour for landing, a stark reminder that they were entering a war zone with all of its inherent risks. As a further precaution, all of the lights were turned off, leaving the cabin in darkness apart from the emergency lighting running along the floor. For those who had never experienced this before it was rather worrying and difficult not to consider the odds of someone trying to bring the aircraft down on its final approach. The landing went smoothly enough, however, and the passengers were able to relax once more as the old Tristar taxied to a halt outside the temporary military terminal.

The ancient city of Kandahar, Afghanistan's second largest conurbation, sits below the foothills of the Hindu Kush at just over 1,000 metres above sea level. As the Grenadiers stepped out into the night air for the first time, they were struck by the unexpectedly cool temperature. The huge ISAF base in Kandahar was home to thousands of soldiers from many nations but these new arrivals were only to spend about eight hours there before moving on. After more briefings they were eventually told to bed themselves down in some basic tented accommodation. Most of them spent a fairly uncomfortable few hours just trying to keep warm in the cold night air. The temperature was close to freezing, which was a shock to many who had arrived prepared for a hot summer tour in the desert. In the morning a hot breakfast was wolfed down before the troops were once again shepherded to another waiting aircraft. They took in the unfamiliar smells and the sight of armed personnel, an early indicator that not very far away there were people who would do them harm given the opportunity.

For the final leg of the journey to Camp Bastion in Helmand, it was necessary to switch to a C-130 Hercules transport aircraft because the runway at Bastion was an improvised one. The flight was mercifully short and the passengers in the crowded cargo bay were relieved when the pot-bellied transport plane began its descent into the Helmand desert. Before long they were able to get their first glance of their new base through the small windows in the fuselage. Camp Bastion was a huge, sprawling, tented camp set in the remote desert landscape. There was little sign of any civilian habitation for several miles around, no water, no roads and no trees, just miles of desert. Perhaps unsurprisingly, the Afghans call this the Dasht-e-Margo, which translates roughly as 'the Desert of Death'.

The Royal Engineers had invested a great deal of time and effort in mine clearance and in bulldozing the red sand to construct a landing strip capable of safely receiving these large transport aircraft on a regular basis. The turboprop planes were required to make a steep descent onto the runway at Bastion where the wash from the propeller blades quickly raised a storm of dust around the vibrating air frame. The C-130 crew efficiently offloaded their passengers and freight. As the Grenadiers walked down the cargo ramp they got their first ground-level view of Bastion. There was a tall perimeter fence with watchtowers placed at regular intervals and an abundance of concrete blast walls placed around various compounds, some of which contained helicopters. Looking out over the runway was an improvised control tower. This was perched precariously on top of a rickety scaffold frame and the whole structure was draped in a very large camouflage net. As far as one could see, in most directions there were tents, hundreds of tents. Some of these were for accommodation and others were stores and operations rooms, the latter could be identified by the myriad of radio antennas which surrounded them. The union flag and various smaller regimental insignia fluttered in the light wind and marked the various unit areas.

Driving through the massive collection of tented administration areas along the dirt roads, they passed a series of helicopter landing pads. These were separated by concrete blast walls and the menacing silhouette of the Apache Longbow attack helicopters could be seen against them. It was clear that other nations were represented here too and those who had served in Iraq were quick to identify Danish vehicles and personnel. There were plenty of US Humvees parked up, their .50 calibre machine guns pointing skyward. The troops were soon placed into some sparse transit tents which had rapidly warmed up in the morning sun. After organising themselves for the following day, the Grenadiers were given the day off to recover from the long journey. Most made themselves comfortable and caught up with some sleep.

The following two days were filled with a mandatory training package known as Reception, Staging, Onward Movement and Integration (RSOI). Whilst everyone had undergone extensive training before deployment, RSOI was designed as a reminder and, more importantly, to bring people up to date with recent events. The local intelligence picture was more current and much more detailed than anything the troops had heard in the UK. They were told about the latest Taliban attacks and the drills that the Royal Marines had developed to defend against them. The distribution of troops around Helmand was of particular interest to the Grenadiers and each man was keen to speculate on which town or village he was likely to be deployed to first. Some places had fierce reputations and the troops had heard such names repeated often in the media. The town of Sangin got special mention because it had been the scene of fierce fighting during the previous summer and the Marines had seen much action there too.

When arriving in an operational theatre, all weapons systems are usually checked for serviceability and accuracy by live firing. This process for small arms is known as zeroing and simply involves moving the sights of the weapon until each shot is hitting the target accurately. It is a simple process but is time-consuming as the procedure has to be repeated for each soldier's personal weapon. Many demonstrations and lectures followed and the Grenadiers were brought up to scratch on the in-theatre procedures for a variety of eventualities.

The training package was soon completed and the troops now looked forward with keen anticipation to joining their advance party in what would be their temporary headquarters for the next six months. The Grenadiers would be leaving Camp Bastion for the nearby forward operating base (FOB) Tombstone. Tombstone was a small American-constructed site tucked away inside a corner of the large Afghan camp, Camp Shorabak. It was so named because the US troops felt its location was in the 'wild west'. Tombstone was separated from the Afghan camp by its own small perimeter and dedicated security force which was generally shared between the US and UK inhabitants. Camp Shorabak was a purpose-built modern barracks designed specifically for the newly formed 3/201 Brigade of the ANA. It was constructed with US dollars and expertise to give the ANA in Helmand a comfortable, secure and modern base from which to operate. Shorabak was constructed on a US-style grid system and consisted of one-storey flat-roofed accommodation buildings, hangars for stores and garages, and a large dining facility. It was surrounded by a high barbed wire fence complete with manned watchtowers. Like Bastion, Shorabak was surrounded by desolate desert and little else. The camp was, however, neat, tidy and very functional, quite different to the crowded mass of tents in Bastion.

The Royal Marines of 45 Commando had organised some continuation training for the newly arrived OMLT group and this involved the opportunity to fire the heavier weapons too. The Grenadiers seized this opportunity to fire the .50 calibre machine gun and the GMG. The latter weapon system was brand new and had only recently been issued in Afghanistan. There had been little opportunity to practise with this highly regarded piece of equipment because few had been available for training in the UK, a situation that was significantly improved for subsequent deployments. The GMG was capable of firing 40mm high explosive grenades accurately at point targets to a range of 1,500 metres; it could do this in bursts and had a rate of fire of up to 340 rounds per minute. It was a potent and deadly weapon and the Grenadiers were keen to hear how it had performed against the Taliban. Hundreds of rounds were fired from the .50 calibre vehicle-mounted machine guns too. During training the few available guns had been shared around the brigade and to the frustration of many the OMLT had not been at the top of the list for their use, and there had also been limited ammunition in the UK. The time was now used to ensure that everyone was absolutely confident on this essential weapon system. It was during the range practices that the Afghan weather once again surprised the new arrivals; there were frequent rain showers which turned the sand to sticky mud and formed large puddles in the numerous tracks. There was little evidence of the baking Afghan heat they had been told about, but few doubted it would arrive soon enough.

Life in FOB Tombstone was refreshingly relaxed after the crowded chaos of Camp Bastion. A series of air-conditioned aluminium huts with cement floors would be home for those who were rear-based at Tombstone. Most of the OMLT would be deployed for long periods, but these buildings would provide some civilisation on their return. Gravel had been laid around the buildings to reduce the amount of dust kicked up and there were temporary showers and washing facilities. The Grenadiers were pleasantly surprised at the standard of their new home; most had served in Iraq and had low expectations.

The American inhabitants who had also recently arrived were a friendly bunch. They were drawn from the US Army, Navy and Air Force: they made up the US Provincial Reconstruction Team and the Logistic Support Team in Helmand, and were commanded by a rather relaxed Lieutenant Colonel of the Airborne Rangers, who was fond of wearing shorts, T-shirts and a baseball cap. His cap was decorated with the badge of the Royal Marines and the 'Grenade Fired Proper' of the Grenadier Guards had recently been added to it. He resembled a sort of cross between an eccentric, two-badged General Montgomery and the easy going Colonel Blake from M*A*S*H. The Grenadiers and their American counterparts soon got to know each other and, although constantly amazed at their cultural differences, the allies got on exceptionally well.

The Grenadiers occupied their accommodation, about 30 men to a building. Each had a simple folding camp bed which was assembled inside a large bell-shaped mosquito net. The accommodation was sparse, but the troops were more than content. The battalion had arrived with its own chefs and under the supervision of the master cook a field kitchen was assembled in a tent behind one of the aluminium structures, which was then utilised as a dining room. Before long these young soldiers were providing the Grenadiers and their CS and CSS colleagues with a standard of food that would be the envy of many UK restaurants. On seeing the UK rations, the American contingent was hugely envious of the standard and choice of fresh food. From the very beginning it became the norm for many of the Americans to be invited into the Grenadier dining room, particularly for Sunday lunch, which they looked forward to. A reciprocal arrangement often took place during which large quantities of luxury US ice cream were consumed. Such home comforts were all very well for those who remained at Tombstone but most would soon be heading out into the desert for very long periods of time.

The serious business of taking over from the Marines continued apace; they had endured a tough six-month tour and were eager to get away. There were large quantities of kit and equipment to be taken over and a certain amount of familiarisation training to be done. Vehicles were a priority, particularly the backbone of OMLT mobility, the WMIK (Weapons Mount Installation Kit). This was a slightly up-armoured and stripped down Land Rover. All non-essential bits and pieces had been removed and some light armour had been fitted as limited protection against mines. A 7.62mm general purpose machine gun (GPMG) was fitted on a swivel in front of the commander. This could be fired from the seated position as there was no windscreen. In the rear there was a circular steel cupola in which the gunner stood and from where he manned the .50 calibre machine gun or '.50 cal' as it was more familiarly known to the troops. The steel side doors had been removed and specially designed lightweight armoured panels replaced them.

All of the vehicles had seen better days and the desert terrain had obviously taken its toll on them. They were battered, worn and every nook and cranny was crammed with red dust. The armour fixed at the doors hung loosely and some of the vehicles had bullet holes in them, evidence of their encounters with the enemy. The Grenadiers were disappointed at the state of the vehicles but were realistic as to the reasons for their poor state. They joked that the whole fleet was held together with masking tape and 'bungees' (the military term for the universal elastic straps used for securing loads). More worrying was the quite small number of serviceable vehicles available for operations. The attrition rate of these vehicles had been reasonably high, some had been destroyed and others had simply given up and seemed determined to defy the best efforts of the REME mechanics who worked tirelessly to get them on the road. There were a few Pinzgauer utility vehicles in a similar condition, which were used for carrying extra ammunition and stores but precious little else. The familiar story of prioritisation was heard and clearly the OMLT was not near the top of the list. This was to become a constant frustration for commanders trying to plan the deployment of their troops. The WMIKs were consequently very precious assets which would need to be taken care of.

Immediately upon arrival, daily meetings or 'O Groups' were held for commanders and key personnel. These meetings would always begin with an update on the intelligence picture and an overview of the day's events. The handover between the two brigades was staged over several weeks in order to allow a gradual changeover on the ground. The outstations in some cases were quite distant and deployment would be achieved by helicopter. It was important too that the incoming troops had time to learn the lay of the land and the local patterns. As the assembled troops listened to the daily events, they became familiar with the main area of operations.

Camp Bastion was a few kilometres to the south of the vital Highway 1 and it was from here that the essential logistic effort was mounted. About 50km to the south-east, the HQ of Task Force Helmand had situated itself in the provincial capital Lashkar Gah. Probably the most important conurbation in Helmand, the town housed the provincial government and was at the heart of the Afghan Development Zone. It was in this highly populated area that ISAF hoped to have the most influence and convince the population that their lives could be improved and peace could be achieved. This message, together with security and reconstruction, was to be pushed up and down the Helmand River valley. Most people lived in or close to the Green Zone, the area on either side of the river's banks, and it had been necessary to secure a number of towns along the valley. The town of Gereshk sat astride Highway 1, about 40km from Bastion and just to the north of Lashkar Gah. A thriving market town and trading centre, Gereshk also provided an important police detachment to guard the small dam and hydro-electric generator on the fast-flowing Helmand River. A large forward operating base had been established to the west of Gereshk and this was home to a UK battlegroup and a number of US personnel; it was known as FOB Price.

A barely usable tarmac and dirt track known as the Route 611 ran north from Gereshk on the eastern side of the Helmand River and was the main traffic artery for a continuous line of small villages sitting close to the river banks. These were almost identical, consisting of countless sand-coloured buildings, each of which was surrounded by a high mud wall, baked solid in the Afghan sun. Route 611 passed straight through the notorious town of Sangin, scene of desperate fighting during the previous summer. UK forces had taken the town centre from the Taliban who ever since had harassed and tried hard to dislodge the British force based there. The 611 was consequently an extremely dangerous route to use. It was frequently mined and any ISAF troops using the road could expect to be attacked. Sangin was nevertheless under Afghan government control and people had begun to return to their homes.

When the route left Sangin it passed a number of small villages and then eventually led on to Kajaki. This was the northern British outpost in Helmand and it was the site of a strategically important dam. The Kajaki Dam was designed to serve a dual purpose. First, it provided power to much of southern Afghanistan and, second, it allowed the irrigation of thousands of acres of farmland downstream. The huge structure, originally built in 1953, stands 100 metres above the valley floor. Years of fighting had ensured that the turbines within the dam operated at less than full capacity. The replacement of the machinery was a key Afghan government and ISAF reconstruction project designed to provide electricity to the long-suffering population. The Taliban, oblivious to the plight of the civilian inhabitants of the south, were set on destroying any such high profile symbol of government control. This meant that a strong British and Afghan force was permanently stationed at Kajaki, some 140km from Bastion, in order to protect the dam.

Following the Helmand River south from Lashkar Gah and about 100km from Camp Bastion, another British and Afghan force had been positioned in the town of Garmsir. This was the southernmost permanent UK outpost in Helmand. Operations to the south of Garmsir were frequently undertaken but there were no other static British troops between this small town and the

Pakistani border, over 200km to the south. Garmsir was consequently seen by the Taliban as the front line; this was the furthest ISAF expansion and they were determined that it should not spread any further south. They had constructed trench systems and defences within range of the UK FOB known as Delhi. Because of the frequent attacks on the base, UK artillery assets had been placed in another isolated desert FOB within range of the Taliban attackers.

Although the task force was responsible for the whole of Helmand, the main area of operations was in effect a 200km-long egg-shaped area, mainly incorporating the Helmand River valley. This area was bordered by the Kajaki Dam in the north and by Garmsir in the south. Inside this strip were several hotspots of enemy activity, notably between Gereshk and Sangin. This river valley would be the scene of much fighting by 12 Brigade in the coming months. The newly arrived Grenadiers were acutely aware that they would be seeing these hotspot soon enough and they continued their preparations for deployment.

Almost immediately the OMLT company commanders and key players in the organisation were introduced to their Afghan counterparts. If the relationship with the Afghans was to succeed, Hatherley would need to win the trust of the 3/205 Brigade Commander, the redoubtable Brigadier General, Muhayadin Gori. A tall, wiry man in his fifties who sported a large black moustache, Muhayadin had been trained by the Soviets and was an experienced soldier and deeply committed to his country and to his men. He could be passionate, and at times frustrated, but underneath lurked a friendly persona which Hatherley went to work on right from the start. It was important to show the general due respect and to listen patiently to his complaints about support from above and the perceived failings of the British Army. Hatherley was exactly the personality required for the job, his archetypal Guards officer's good manners and patient demeanour were exactly what was needed to forge an effective working relationship. Muhayadin sometimes disliked the news that Hatherley brought him, but when put the right way he accepted the realities and frustrations of working inside a coalition and would often break out into a toothy grin as though he had expected the worst all along and the situation was actually no surprise to him. Hatherley had a tough job in mentoring this old war horse and it would require every piece of patience he possessed to prise the Afghan away from the inflexible Soviet doctrine that he had been taught. All of the senior Afghan officers had undergone training by the Americans at the ANA Corps HQ near Kandahar. The mixture of Soviet and US teaching combined with the more recent British attempts at military education must have been quite trying for Muhayadin, but as the British were about to discover the Afghans had an amazing capacity to learn lessons and then to completely ignore them.

Muhayadin's brigade was widely spread around Helmand and was usually to be found in all of the FOBs and outposts inhabited by the British. They were under strength and widely dispersed. Few if any of the ANA soldiers were Helmandi, they were drawn from other provinces of the country, mainly those around Kabul. Their predominant language was Dari and most hated the Taliban. In principle the brigade operated a three-way rotation with one kandak being deployed in the field, one on leave and the other engaged in training with the OMLT at Camp Shorabak. In practice this had been impossible to achieve because of the tempo of operations. Most had spent far too long in the field, nearly all were overdue leave and were owed money that had not reached them when deployed. Very little meaningful training had been conducted and much of it had been done 'on the job' when deployed. A large number of soldiers were listed as absent without leave (AWOL). This was worrying and was a possible indicator of poor morale among the Afghan troops. It was soon pointed out, however, that this was quite normal. Afghan soldiers were usually flown to

Kandahar or Kabul to start their journeys home to their families, but most had hundreds of kilometres still to cover after they had reached these cities. With no rail or public transport network and very poor quality roads, it was a major feat for some even to reach home, let alone to return to camp on time. Amazingly, most did return, but leave dates were often quite meaningless.

It was obviously going to take some time for the Grenadiers to understand the ways of the ANA. The mindset of a Guardsman is to carry out instructions to the highest standard, accurately and immediately. This was going to be at odds with the Afghan way of life, which was more laid back. There would have to be some compromise and it was difficult to see how the Afghans would change their ways. This was, after all, their country. They had fought the Taliban for years, had lost many of their friends and would still be here fighting when the Grenadiers had returned to the UK. A great deal of patience, understanding and at times robust leadership from the British was going to be needed. Some of the Grenadier Guards mentoring team were only days away from their first encounters with the enemy and they would shortly learn what the Afghans had been doing for years.

4

PUTTING OUT FEELERS

For the Grenadiers of 2 Company, the stay in FOB Tombstone was a brief one. Only two days after completing their training and only five days after landing in Kandahar, they drove out of the base in order to locate and join their Afghan charges in the 2nd Kandak. This task was complicated by the fact that they were widely spread out and in different towns. Each OMLT platoon would mentor one Afghan company. The problem was that an OMLT platoon was nowhere near normal platoon size and in reality consisted of about ten men. One party, comprised of some WMIK vehicles and commanded by Lieutenant Rupert Stevens, headed to FOB Price and to the nearby town of Gereshk. This was the first time that a Grenadier patrol had moved any distance from Tombstone and it was a tense initiation. The drivers soon learned that driving in the desert was very hard work and took great concentration. Most tried to steer their vehicles into the tyre tracks of the vehicle in front to reduce the possibility of hitting a buried mine, of which there were many. It was well known that the Taliban often dug up 'legacy mines' left by the retreating Soviets and placed them where they knew coalition vehicles would drive. It was therefore important to avoid setting patterns. The commanders seated next to the drivers kept one hand on the plastic butt of their GPMGs or 'Gympies' in case they were needed. The gunners standing in their cupolas each held a pack of small coloured flares. These were used to warn any oncoming traffic to pull over or to stay away. The risk from vehicle-delivered suicide bombs was extremely high and this non-lethal and highly visible method of warning people was one of a number of escalating measures leading ultimately to the use of lethal force if it was necessary. The decision to open fire at an oncoming vehicle would be a split second one and everyone was aware of just how difficult this might be. The threat from suicide bombers here was much greater than during the previous summer in Iraq and any civilian who came too close could be a potential bomber. All of this made for tense and tiring work.

FOB Price was quite a sight. Located just off the main highway it could be identified by a tall steel watchtower which dominated the area. A roughly circular perimeter had been constructed by sand-filled bastion walls and much barbed wire. It was vaguely reminiscent of the US fire bases from the Vietnam war. The interior was relatively civilised and the UK Battlegroup Centre were in the process of taking over from the Royal Marines. A significant number of US troops also occupied the dusty interior of FOB Price and their Humvees were present in large numbers. A short distance outside the perimeter a second, smaller compound had been constructed. This was the ANA compound which had recently been built; there were several tin-roofed buildings, an administration area and little else. Like their colleagues in Camp Shorabak, the Grenadiers of 5 Platoon now had to meet their Afghan counterparts. They were soon to discover the Afghan obsession with tea drinking. The Afghan brew was known as chai and was taken without milk. It was new to the British and the yellowy liquid was definitely an acquired taste. Chai was always drunk during meetings and social get-togethers. Normally poured into small glasses from large copper kettles by an orderly, chai was consumed in great quantities.

Stevens and his NCOs sat cross-legged on the floor and tried to break the ice with their charges. In the coming days they discovered some other Afghan idiosyncrasies, which were quite at odds with the ways of the British Army and the Grenadier Guards in particular. The ANA soldiers were very fond of wearing brightly coloured bandanas around their heads and necks. These garish scarves often replaced the steel helmets that should have been worn on patrol. They were also fond of decorating their rifles with colourful stickers and even painted flowers. It was not at all unusual to find flowers protruding from the barrels of the soldiers' AK-47 rifles. The Grenadier NCOs were quite bemused by this behaviour and early attempts to guide the Afghans away from these practices were soon abandoned. Another difference was that the standard Afghan patrol vehicle was the Ford Ranger pickup truck. The rangers were not military vehicles at all but had been provided in large numbers by the Americans along with some huge 6 x 6 utility trucks. The vehicles were sprayed a sand colour and a large ANA emblem was stuck on the doors. There was absolutely no armour or enhanced protection on these vehicles and their off-road manoeuvrability was average at best.

Patrol preparations by the British normally took some time. The vehicles were inspected for serviceability and all equipment was checked and stowed in a specific place on each WMIK. Detailed orders were given and all likely 'actions on' – the planned reactions to various scenarios – were briefed to the patrol members. All weapons were checked and checked again. By contrast the Afghans clambered aboard their Ford Rangers in a seemingly haphazard fashion. The soldiers sat on the sides of the pickups festooned with ammunition belts, toting their brightly decorated weapons, cigarettes dangling from their mouths and raced off into the desert at a ridiculous speed, which the British struggled to safely match. Visible through the clouds of ensuing dust were the pointed tips of the RPG rockets which were stored in makeshift cylinders behind the cabs of the vehicles. They protruded like porcupine spines and looked distinctly unsafe. It wasn't long before joint patrols were regularly taking place in the Gereshk area and 5 Platoon soon established a patrol routine with the Afghans.

Movement through the town, with its busy main street, was always a concern. All manner of vehicles came in from the main highway and hundreds of civilians making their way to and from the bazaar mingled at the sides of the road. Any one of them was a potential suicide bomber so there was always great relief when the patrol entered the gates of the district centre which was controlled by the ANA and provided a relatively safe haven. There was one early minor success for the ANA accompanying 5 Platoon. A motorbike was stolen at gunpoint and the ANA men acted quickly and were able to apprehend the thief. There was some excitement at the time, but the perpetrator appeared to have been working alone. He was handed over to the police and that was the last that 5 Platoon heard of the man.

Further to the north, 4 Platoon had driven to FOB Robinson, near Sangin, more commonly referred to as FOB Rob. Here they shared the same living accommodation as their Afghan charges and a similar relationship had been forged as that achieved by 5 Platoon. Captain Ed Janvrin, an attached British Gurkha officer, was in temporary command and Lance Sergeant Owen soon got to grips with the administration, trying to gently guide the Afghans into adopting British military practices for patrol preparation. Logistics were rarely even considered; the Afghans were used to living off food from the scattered villages where they could purchase bread and fresh meat for cooking, which was always done over open fires. Large quantities of US MREs (meals ready to eat), the US field rations, were provided for the long patrols but these were very unpopular with the Afghans who looked at the pressure- packed plastic blocks with distaste. By contrast the Afghan meals usually consisted of communally cooked vats of curry and rice with pita bread and, surprisingly, cans of cold Coca-Cola. The Grenadiers

were often invited to join the Afghans at meal times and the officers
were usually keen to do so. This was an ideal opportunity to build
trust and understanding with the ANA officers, most of whom had
been in the field for long periods of time. Some of the more junior
British soldiers were less keen to join these bonding sessions, not
because of a reluctance to engage with the Afghans but through a
fear of the food. Most had seen the meat hanging in the bazaars,
covered in flies and dripping blood. Good hygiene was something
most had learned in Iraq and they had no wish to develop diarrhoea
or a vomiting illness which is no joke when you are living in the
desert with no decent latrines or showers for miles.

Patrolling in the Sangin area was a potentially deadly pastime. The
Taliban were very active in the Helmand River valley between
Sangin and Gereshk. They were intent on harassing the British and
Afghan government troops at every opportunity. Sangin was a
town that they were keen to recapture and they regularly mounted
ambushes and attacks. The British troops had significantly more
firepower available to them, and in the many gun battles the
Taliban usually came off worst. FOB Rob contained 105mm guns
and these could deliver accurate fire on call. Additionally, Apache
attack helicopters could be called forward and fast jets were usually
quickly overhead to drop bombs onto any potential attacker. This
impressive array of weaponry had not gone unnoticed by the
enemy who had lost hundreds of fighters in the area. Unfortunately
for the men of 12 Mechanised Brigade, the Taliban would alter
their tactics over the coming months. They would come to rely
more heavily on mining the routes, especially the notorious 611.
They had learned lessons from Iraq too and roadside bombs would
also become the norm in Helmand.

The conditions in Kajaki, Task Force Helmand's most northerly
outpost, were similarly sparse. It was to here that 6 Platoon
had been delivered by RAF Chinook helicopter. The journey by

Chinook, the 'workhorse' of the army in Helmand, was crowded and noisy. The troops were able to look down on the desert wilderness below them. For miles there was nothing but sand until the Helmand River came into view. Then civilisation in the form of hundreds of identical brown buildings, each standing in its own walled compound, could be seen, and these became more apparent as the river got closer. Kajaki nestles in the foothills of the Hindu Kush and the impressive mountains were plainly on view as the twin-rotor helicopter deposited its troops. Second Lieutenant Howard Cordle and Sergeant Ty-lee Bearder together with the remainder of 6 Platoon were now living with their Afghans in the shells of some ruined houses. The views were breathtaking and the deep blue of the Helmand River's waters seemed to change colour during the day. Each morning the British troops woke to the sound of the river flowing into the Helmand valley below them. The OMLT here were located with another company of UK troops from the Royal Anglians, whose job it was to secure the vital dam and to provide security to the local population. The soldiers were essential to the latter task and the Grenadiers joined their charges for daily clearance patrols into the surrounding villages. It was important for the civilian population to see Afghan government troops on patrol.

The surrounding hills and villages were alive with Taliban fighters. The British lacked the combat power to push out and completely dominate the area, although frequent clearance operations were mounted. These operations were invariably successful but, unfortunately, as soon as the British and ANA troops left the area, the Taliban were able to move back in. Their attrition rate was extremely high but they always seemed to be able to regenerate.

The town of Musa Qala, a Taliban stronghold, was a just across the mountains to the west. The previous summer, the provincial governor had brokered an agreement with the tribal elders to establish an exclusion zone around the town, which ISAF troops would not enter in return for tribal elders denying Taliban forces access. In this way the town could be left to get on with life without the belligerents. After 143 days the agreement broke down when the Taliban went back on the deal and reoccupied the town left vacant by the British. Musa Qala was now a Taliban stronghold once again and it was to here that the enemy fighters often went to lick their wounds and to get further direction. Coalition forces operated in the surrounding countryside and were able to make life difficult for the Taliban, but the close proximity of Musa Qala to both Kajaki and Sangin meant that the enemy was able to slip through and attack troops in both areas.

The last part of 2 Company was the Fire Support Platoon under Captain James Fox. 'Foxy' and his small group of support weapons specialists were sent to Lashkar Gah. The ANA compound here was located very close to the river and was a stone's throw from the HQ of Task Force Helmand. Patrolling in the busy city centre was again a nervy business but the Grenadiers quickly realised that the ANA soldiers had an awesome ability to spot things that were out of the ordinary. They were also able to engage with the local population and the Grenadiers found that the local people seemed to be more relaxed around them when the ANA were present. 'Lash' was more secure than the towns to the north; there was a large ANA and police presence here in addition to the many coalition troops that patrolled the area. Reconstruction and development was very visible, the town was thriving and Taliban influence were less significant than in other parts of Helmand. They were still able to mount attacks but these were fewer in this area. The Grenadiers could sense progress in Lashkar Gah and the people here had hope for the future. The small 2 Company HQ busied itself in doing its best to support the deployed platoons, widely dispersed as they were. Daily reports were received in Shorabak from each of the locations and inevitably 'shopping lists' of equipment would be handed to the harassed company quartermaster sergeant who did his best to meet the demands of the deployed troops. In meeting these demands, everyone at Shorabak was well aware of the harsh conditions in which people were operating to the north.

Major Barnes-Taylor, the Commander of 2 Company, and his HQ visited as many locations as they could but Kajaki was always a difficult place to get to because the only practical way in was by Chinook. These precious assets were in great demand and Task Force Helmand had very few of them; consequently passengers and cargos were closely scrutinised and prioritised.

Even as the main body of 2 Company deployed into its widely dispersed area of responsibility, one of its number was already in contact with the Taliban. T had deployed with the Royal Marines on one of their final operations in the Kajaki area. They were short of a sniper and he fitted the bill exactly. The Marines had planned a sniper ambush in an area where intelligence suggested that the enemy might expose himself. The plan was for four sniper pairs to infiltrate the area and to conceal themselves. Each pair was equipped with a .338 long-range rifle. T was equipped with an L96 7.62mm sniper rifle, which had a much reduced range compared with the .338. He was consequently deployed to a blocking position close to the Marines' company HQ. Six months of experience had taught the Marines that when they contacted the Taliban, the enemy usually tried to get around the flanks of their attackers. It was not unusual to find the enemy attacking you not only from an unexpected direction but also from the rear.

The ambush patrol left the security of the Kajaki FOB at 0330 hours. The last of the winter chill was in the air although it soon disappeared as the sun came up. Despite the cool of the early morning, the troops were drenched in sweat by the time they arrived in their pre-designated ambush positions. This was T's first patrol; the other Grenadiers had not yet fully taken over from their Marine counterparts so he was on his own. He was conscious that he was the only man on this patrol who had not yet seen action and he tried hard not to show his nervousness. He and his partner positioned themselves on top of a flat-roofed building. Lying there they could observe the vegetation of the Green Zone.

The desert sun rose steadily throughout the morning and there was no sign of the Taliban. By early afternoon everyone was feeling very tired in the now scorching heat. Suddenly the radios crackled into life and T listened to a sighting report of suspected enemy activity coming from the area of the ambush. There was a positive identification (PID), of the enemy and shortly afterwards gunfire could be heard coming from the left. It was a good way off, but by the sound of the radio several Taliban fighters would not be going home for dinner on this particular evening. T and his spotter continued to scan the distance for signs of movement and a short time later they were rewarded by the sight of three enemy fighters in the distance. One of them was carrying a Russian-made rocket propelled grenade (RPG) launcher. Their baggy clothes could be seen blowing in the wind and T realised that the wind conditions would complicate the shot. The spotter estimated the range to the target as being about 1,000 metres and T agreed. This was a long way off, but he adjusted his sights to the estimated range.

As the two men watched, the enemy fighters got down behind a bank. Their heads could be seen popping up and they appeared to be observing for likely targets. T was sweating now, this was the first time he had fired at a live target. He struggled to get a decent sight picture through his scope, as his helmet made this rather awkward. He quickly unfastened the chin strap and threw his helmet aside. He pushed off the safety catch and took careful aim at the first of the fighters. The target was smaller now but T could clearly see the man's head and shoulders above the bank. He controlled his breathing and slowly squeezed the trigger. The spotter observed through binoculars and T attempted to follow through his shot before working the bolt and chambering a second 7.62mm round. The shot missed and T swore. The spotter offered some observations on where the bullet had struck and the young sniper adjusted his sights before acquiring his target again. Remarkably, the Taliban fighter had once again presented himself above the cover. T fired once more and again the round failed to find its target. He adjusted the sights, conscious that the Marines behind were watching. At 1,000 metres this was a very difficult shot, one which was much better suited to the .338 snipers. Once more the enemy fighter raised himself up as a target. T squeezed the trigger again as the sweat dripped off his chin and this time the 7.62mm bullet hit its target. Even a kilometre away both the Marine and T saw the man go down, he had definitely been hit. The sniper worked the bolt once again but this time the enemy kept their heads down. He had their range now and if anyone else was stupid enough to show himself T would bag him too.

It was clear that the Taliban were infiltrating the area and fire was now being taken by several of the British positions. The sniper pair were now ordered to withdraw to the position of the company HQ. As they rose from the roof, the air cracked with the sound of bullets. The two clambered down as quickly as possible, conscious that fire was being directed at them from several different positions. It was now necessary to cross some open ground in order to join a drainage ditch which they could use as cover. The Marine threw a smoke grenade, there was a loud 'pop' and a thick white cloud of phosphorous smoke engulfed the area. The two sprinted as hard as they could for the ditch as their trail was traced by the characteristic crack of a PKM machine gun. They dropped into the ditch and continued to run in a crouched position. T was excited and terrified all at the same time but the danger seemed to be receding at least. The PKM was still firing overhead but the pair felt relatively safe in the ditch.

Meanwhile the Marines had called in air strikes and the roar of jet engines could be heard as the ground behind them shuddered under the impact of several bombs. Brown smoke emerged from the destroyed compounds, which seconds earlier had been enemy fire positions. This was enough to make the Taliban break contact, at least temporarily. Once they reached safety, T was greeted by grinning Marines who teased him for taking three shots to hit the man. They were entitled to be light-hearted; they were going home in a matter of days.

There was a long hard march back to Kajaki. Everyone dripped with sweat and panted heavily under the weight of their equipment and the heat of the day. Their heavy body armour, while essential, restricted their movement and caused them to sweat even more. Their shirts were drenched and, once he had got back, T was exhausted by the effort. There was great interest in his adventures from the other Grenadiers and he was forced to recount the story over and over again. It didn't dawn on any of them that in seven months' time this encounter would be so insignificant as to be hardly remembered. There would soon be plenty of action and T would be in the thick of it.

The Inkerman Company too had been busy. 3rd Kandak were currently deployed with their Royal Marines mentors and it would be necessary to conduct a handover in the field. 'The Ribs',* as the Inkerman Company were known, had trained hard since their arrival and now felt that they were ready. There had

---

* A term originally applied to 3rd Battalion Grenadier Guards who served as Marines in the eighteenth century and were accommodated in the 'ribs' of the ships. The Inkerman Company maintains the customs and traditions of the 3rd Battalion.

been recent Taliban activity in the area to the north of Lashkar Gah on the edge of the Afghan Development Zone. Operation Malachite had been mounted by the Marines to clear the area between Lashkar Gah and Gereshk, south of the main highway. Major Marcus Elliot-Square now had the unenviable task of taking over responsibility for the 3rd Kandak and for ensuring that any remaining Taliban were kept out of the area. 3rd Kandak were centred on the village of Babaji and it was to here that the Ribs headed in early April to relieve the Marines. The Inkerman Company moved in a convoy of WMIKs and Pinzgauers. They conducted a resupply for the ANA troops at their small camp near FOB Price and then headed south to Babaji. The distances were not great and the company arrived at their rendezvous on the banks of the Helmand River intact after only a couple of hours. The plan was for the Grenadiers to shadow the Marines, meet their ANA charges and get a feel for the ground. Once a detailed handover was completed, the newly arrived OMLT troops would take the reins.

Elliot-Square's head was full of questions for the Royal Marine company commander whom he was relieving. There was clearly much to learn about the area, the operation and of course about 3rd Kandak and its various personalities. Unfortunately for the Inkerman Company things had changed. The handover date between the Marines and the Grenadier OMLT had been brought forward by four days. The flights to get the Marines out of Helmand were booked and they had very little time to get themselves back to Tombstone in time to carry out the necessary administration and to make their flights out. As a consequence the handover was extremely brief. The two officers discussed a few points, Elliot-Square got in as many questions as he could; the Marine handed over a map of the area and wished him good luck. A short time later, the eager Marines drove off in the direction of Camp Bastion and were soon nothing more than a dust trail. They looked around at the rag-tag army they had inherited and wondered where to start.

The Afghans seemed unconcerned that their original mentors had deserted them and that another group of worried-looking British soldiers had taken their place. By now the sight of their bandanas and flowers had become quite normal. The 3rd Kandak looked battle-hardened and so they should have, many were ex-Mujahedeen fighters. They were deeply tanned and many wore thick beards like their commander Colonel Rassoul. As the 3rd Kandak commander, Rassoul had control over all aspects of ANA business. Elliot-Square introduced himself with a firm handshake and was rewarded with a warm smile. Through an interpreter he was able to break the ice and the two were soon discussing business over some chai. Rassoul had been a Mujahedeen commander and was deeply respected by his men. In his late forties, he had seen much action. His unit had recently been fighting in Kandahar Province; they had spent a very long time in the field and were quite weary. It was clear from the outset that this bearded warrior was very committed and capable. He exuded quiet confidence and was dedicated to the cause of his government and to the future of his people.

The Marines had already supervised the ANA on a clearance through the area but a number of compounds had been by-passed. The first task for the Inkerman men together with their Afghan colleagues was to check these compounds for the Taliban. The joint ANA and OMLT group set up a temporary leaguer between some compounds and from here they sent out regular patrols to dominate the area and to deter the enemy from reoccupying the areas that they had been driven from. The ANA established several small outposts overlooking crossing points on the river and these were regularly attacked after dark. The Taliban were driven off each time but this demonstrated their determination to retain influence in the area now occupied by coalition forces.

The scenery along the banks of the Helmand River was quite beautiful and the poppies were in full bloom. The landscape would have been an ideal place for a picnic or a quiet walk were it not for the unseen danger that lurked all around. In the middle of all this danger lived the local civilians. The Inkerman Company troops had so far had very little contact with the villagers, so they were interested to see what their reactions would be. During the first couple of days of the Babaji deployment the plight of the Helmand civilians was brought starkly home to the British troops. On entering a small village close to the river, Elliot-Square and his troops encountered two Afghan teachers. The men were clearly pleased to see the British but their fear was palpable. Their nervousness was due to a recent Taliban visit to the village; one of their friends, also a teacher, had been murdered by a ruthless armed group. This was a typical Taliban tactic. By slaughtering anyone who showed signs of intelligence or whose opinions differed in any way from their own, they were able to terrorise the rural population.

The two teachers were in a terrible quandary; although they were in fear of their own lives in case they should be seen by the Taliban, they desperately needed help from the Grenadiers. As they moved further into the village, Elliot-Square and his men soon discovered why. During the previous evening heavy rains further upriver had caused the Helmand River to flood and burst its banks. A 1.5-metre-high wall of water had swept through the village causing devastation. Weaker structures had been destroyed and the powerful waters had swept many people away. Young children and old people had simply disappeared in the confusion. The poppy fields in this area were now covered in a layer of silt 12 inches deep, adding to the villagers' problems. There were many people missing and many of the survivors were still in shock. Elliot-Square surveyed the devastation and talked to the teachers to see what help could be provided. Despite their terrible plight, the villagers were able to sit down with the British troops and to share their own bread, milk and the inevitable chai. The British officer was able to use his radio to contact the Province Reconstruction Team and the civil affairs representatives who deployed to the village in due course. In this instance real humanitarian aid was delivered to the needy villagers, but those who asked the British for help had risked their lives to do so. This was an early opportunity for the Inkerman Company to see the human face of Afghanistan; help was so badly needed in these rural communities where the harsh environment made existence difficult.

The Inkerman Company soon got to know their ANA charges, their idiosyncrasies, habits, strengths and weaknesses. Like 2 Company, they quickly established an effective working relationship. When the joint OMLT and ANA patrols entered the various villages it was common for Rassoul to order that the inhabitants be gathered so that he could address them. Large crowds of villagers would then be assembled and Rassoul, standing on a piece of raised ground, would introduce himself to them. He would explain that he represented their government and that the government was trying to achieve reconstruction and development in order to improve their lives. He warned them against cooperating with the Taliban and encouraged them to report any suspicious movements. It was clear to the British onlookers that Rassoul had tremendous presence and that he was able to get through to these local folk in a way that western troops could never hope to achieve. It was also obvious that he, like other ANA officers, had contacts all over Helmand who were able to pass information to him by mobile phone. In the months ahead the OMLT troops would be able to see first hand that 3rd Kandak were able to act on 'hot' intelligence gained by Rassoul and his soldiers, unhindered by western military restrictions. Some impressive results were achieved through intelligence gained by the ANA. Perhaps more importantly, the information passed on by peasant farmers was used to identify threats and imminent attacks which were planned against both ANA and British troops. There can be no doubt that many lives were saved in this way. Elliot-Square and his team attended shuras or meetings with the tribal and village elders almost daily. They witnessed Rassoul's skill in negotiating with the civilian leaders. His presence provided confidence to the villagers, most of whom were no friend of the Taliban.

After about a week, things were going very well. The Inkerman OMLT had had no serious trouble from the enemy and although the ANA outposts were still being attacked, this was assessed as harassment rather than full-on assault. Rassoul announced that he had some specific information about some Taliban still in the area. The Afghan officer decided that he would act on this information quickly. Elliot-Square advised caution in case of a 'come on'. The British Army had learned many hard lessons from years in Northern Ireland; terrorists often set them up for attack with false information, so the British officer was suspicious. The Afghans were much more impulsive and soon Rassoul was heading off to check out the information. The Inkerman Company had little choice but to accompany their enthusiastic charges.

On arrival at the suspect location, the joint ANA and OMLT force occupied a large compound so that they could fan out and clear the area systematically. Sergeant Dan Moore was bringing up the rear of the convoy with Captain Alex Corbet-Burcher who was in command of the small reserve force. The packet of vehicles closed up along an adjacent compound wall and prepared to cross a wadi to the main force. As they did so they suddenly received a warning from the main force across the wadi that there was movement on the roof of the building that they were now parked up against. Moore ordered his gunner to stand on the rear cupola of the WMIK and to look over the wall onto the roof. As he did so the gunner shouted a warning. A man had indeed been observing the force across the wadi and seeing the head of the gunner pop over the roof, he had panicked and fled. Corbet-Burcher now ordered a clearance of the compound behind the mud wall. The small group of Grenadiers accompanied by the ANA men moved into the compound and began a systematic and cautious search. The whole compound had been prepared for defence by the Taliban. It was a rabbit warren of trenches connecting to the interior of the buildings. Holes had been knocked through walls and firing slits had been constructed. There were tunnels leading who knows where and a deep hole had been dug inside one of the buildings.

As the clearance continued there was a short burst of automatic fire; it came from the far side of the compound in the area of a large poppy field. The ANA soldiers were quick to respond and automatic fire was returned from all over the place. Two Taliban fighters were seen briefly running away from the fields and the ANA contingent raced off in pursuit. Large quantities of ammunition were unleashed in the direction of the running enemy fighters and one ANA soldier even loosed off an RPG rocket as he was chasing them. Predictably it arced skywards and had no effect on the enemy. Corbet-Burcher was now concerned that the ANA men might be sucked into an ambush in the close country and he called them back. It took some time for the shooting to stop but Moore eventually managed to bring the ANA back under control. This had been the Inkerman Company's first look at the 3rd Kandak under fire. They had responded quickly, aggressively and, while their approach was not particularly subtle, it was reassuring to know that they could be relied upon when in contact. Controlling them would be a whole different ball game. Several follow-up operations were carried out over the next few days but the enemy decided to keep a low profile.

After about ten days in the field Colonel Rassoul expressed a wish to enter Babaji on market day. He wanted the chance to address the local people when a good number would be in the village. The civilians had been given a very hard time by the Taliban and now that the enemy had been cleared out of the area, Rassoul wanted to provide some reassurance and to gather intelligence. Elliot-Square decided to accompany Rassoul and the Grenadiers set up a secure perimeter to allow the ANA to move into the centre of town. The Afghan commander addressed the assembled masses for about 45 minutes, his speech seemed to be well received by the people and the perimeter was collapsed in order that the men might return to the joint leaguer. The Grenadiers had been in position for about an hour and a half and the whole thing had passed off without incident. As the convoy moved away from the village there was a tremendous explosion which seemed to have come from the area of the market. Black smoke was seen drifting away and the troops decided to return in order to investigate. They were quickly directed to a large crater. The police and the ANA men immediately started to round up any likely witnesses. No one had been injured but the crater suggested a very substantial roadside bomb. It had clearly been intended to hit the coalition troops; in fact, Colour Sergeant Bastin's WMIK had driven past this very spot about two minutes before the detonation. It was a mystery why it hadn't detonated earlier. This lucky escape was a warning to the Grenadiers and to the remainder of 12 Brigade that the Taliban were beginning to favour the roadside bomb.

5

HERDING CATS

There had been no action for the remaining OMLT companies but like the Inkerman and 2 Companies they set about trying to build a relationship with their Afghan counterparts. The CSS teams had the task of trying to kick-start the Afghan logistics chain which was woefully inefficient. For experienced soldiers who were professional logisticians, mechanics and medics this was a nightmare. The Afghans had few systems in place and there was little or no understanding of the principles to be applied in order to create some form of workable administration. The brigade's vehicles were held by the individual kandaks and it was clear that there was no maintenance system or safety checks and the competence of some of the drivers was clearly in question. Behind the large workshop building was a car park full of wrecked Ford Rangers and large trucks. Some of these had obviously been the victims of Taliban activity but most were the result of accidents. This veritable scrap yard was being used for spare parts and the cannibalised hulks lay all over the place. This practice had to be stopped so that a proper system of prioritisation for repair could be put in place. The number of accidents related to poor driving needed to be minimised too. The ANA had suffered a number of fatal casualties through traffic accidents and it was clear that there needed to be more discipline and some education on the perils of speeding and recklessness. The vehicles were quickly organised into a pool and were allocated to the kandaks as required. Captain Dave Groom, the Grenadiers' transport officer, was instrumental in providing some initial organisation to the transport section. Vehicles just disappeared from the area of the Afghan motor pool during a variety of tasks. Not a man to be thwarted, Groom drew together an Afghan work party and started to construct a barbed wire perimeter fence. This was unpopular work for the Afghans and they gradually disappeared leaving the furious Grenadier officer with only a handful of UK soldiers to complete the backbreaking work under the Afghan sun. The job was nevertheless finished, leaving the ANA vehicles inside a secure perimeter. Any vehicle leaving the small compound had to pass by a sentry and this greatly helped to reduce the attrition rate on the vehicles.

There was a similar story when it came to fuel accounting and tight controls were put in place by the UK CSS troops here too. At all stages the Afghan chain of command was involved in these changes and supported them; the improvements were plain to see, although they were not popular with everyone. The adoption of these procedures was a culture shock to the Afghan soldiers who genuinely tried to understand the changes, but for some the rationale was just too alien to grasp. The mentoring troops were often taken aback by the scale of the task facing them. The ANA weapons were in a similar state to the vehicles and there were a large number of heavy machine guns awaiting repair, there was a shortage of spare parts and no one could find any oil suitable for the weapons. The Infantry OMLTs soon discovered that it was common practice to use diesel as a lubricant on the AK-47 rifles, a practice that was not acceptable to the Grenadiers. There were no Afghan armourers available and remarkably it was usual for the military imam, the brigade holy man to repair the machine guns. The heavily bearded cleric did a very good job in most cases but the REME soldiers were soon running courses on weapons maintenance and mechanics. The latter was very popular with the Afghans and they showed a tremendous ability to learn and to improvise with scant resources. The ANA repair workshops were soon working flat out under OMLT supervision and the enthusiasm and rapid progress was very satisfying for those involved. It was frustrating, however, that the logistic chain went straight to the American-supervised Afghan Corps HQ. There was little that the British could do to influence some critical shortages other than to repeatedly raise the issue at corps level. All of the ANA training was run according to American doctrine. US military contractors delivered much of the high level and logistic training to the ANA and although it was professionally delivered, it remained challenging to break through Afghan culture in order to progress. Many of the Afghan soldiers had been fighting the Taliban for years in one capacity or another and the need for change was not always fully understood at the outset.

For the other companies there had been fewer troops available. Most of 1st Kandak were on leave and much of the remainder of the brigade were deployed. Colonel Rassoul's recently returned 3rd Kandak were despatched on leave by the Inkerman Company. The Ribs had the almost impossible task of coordinating the flights for hundreds of Afghan soldiers so that they could get out of Helmand and closer to their homes. For the RAF the task was relatively straightforward, although the availability of aircraft was always a problem. For the Inkerman Company the seemingly clear-cut task was a nightmare. Just assembling a simple list of names of those who would be flying was a real challenge. The Afghan concept of time is quite different to the British understanding and it proved virtually impossible to get the right people in the right place at the correct time to fly. Somehow the eager Afghans were herded onto the C-130s with some Grenadier escorts and as the aircraft soared into the distance the Ribs were left wondering if they would ever see their charges again.

The other OMLT troops set about the task of training the 1st Kandak and the CS and CSS troops left in Shorabak. This was in addition to their UK continuation training. Ranges were run and the Grenadiers were keen to see how proficient the Afghans were with their AK-47 rifles and PKM light machine guns. Their worst fears were soon confirmed. The standard of marksmanship was generally very poor. This was undoubtedly due in part to the very poor state of the ageing AKs; there were several different types, although they all conformed to the basic Kalashnikov design and the parts were interchangeable. These weapons were designed primarily to be mass produced, rather than for accuracy; they were robust and generally reliable but the diesel oil lubricant did little to help. Large quantities of ammunition were blasted off at a rapid rate with little impact on the target. There was no concept of zeroing or of one man being responsible for his own weapon. The Grenadiers spent a great deal of time teaching the Afghans how to shoot and to alter the sights of the weapon to improve the accuracy to the individual firer. This was usually to no avail because the following day the Afghan soldiers appeared with completely different weapons, rendering all of the previous day's work pointless. Despite these frustrations, progress was made and in later tours newer US-made weapons replaced the tired Soviet-era rifles and improved capability significantly.

It was also necessary to look at how weapons and ammunition were accounted for in the kandaks. The first step would be to find where the weapons were stored and what maintenance procedures were in place. By now expectations were relatively low, but the quartermaster sergeant (CQMS) of the Queen's Company and his work party were amazed when they opened the door of the building being used as a store by 1st Kandak. As the door was pushed open, weapons literally started to fall out and had to be caught by the Grenadiers. Inside was a mass of military equipment which appeared to have been thrown into the room and the door rapidly shut as the guilty parties left. In addition to weapons and miscellaneous stores items, there were ammunition boxes and, worryingly, items of live ammunition, which seemed to have been discarded.

The Inkerman Company encountered an identical problem and the building housing the 3rd Kandak general stores was no better. Items of uniform and military clothing which were expected to be stacked on shelves were lying in heaps on the floor in a dreadful condition. Clearing this mess up was going to take some time and would be slower as the Afghans didn't seem to understand that there was a problem. The ANA NCOs were, however, keen to learn and to please their mentors In order to account for the weapons, every single item had to be removed from the building now termed an 'armoury'. The weapons were sorted into piles with the help of the Afghan soldiers and anything that was in a dangerous or unusable condition was set to one side. As the troops worked their way through the piles of equipment and boxes, hand grenades and RPG missiles were found in dangerous numbers; these were removed and the Grenadier NCOs explained to the ANA soldiers why these practices were unsafe. It was obvious that the ANA found it quite acceptable to leave loaded weapons in the armoury. These habits were indicative of an army that was used to having its loaded weapons to hand; there was little concept of the need for the peacetime procedures that were so familiar to the British. It was going to be necessary for the Grenadiers to effect culture change, but it was clear that compromise was required. The Afghans had immense combat experience and that deserved respect. The Grenadiers were keen to lead by example and to demonstrate how things were done and how quickly such problems could be solved by good soldiers. Progress was slow and much of the initial work was done by the Grenadiers themselves. Once the Afghans started to see results, they could then understand the logic and they enthusiastically helped to clean and to arrange the weapons and ammunition.

RSM Andrew 'Stumpy' Keeley had been active in establishing various working procedures and standing practices in FOB Tombstone. These ranged from security to duties, and extended to matters such as orders of dress. In true Grenadier fashion there were no grey areas left for the troops in Tombstone. Keeley was also to be responsible for mentoring the Afghan brigade sergeant major, a task that would not be close to the heart of any Foot Guards sergeant major. Stumpy's counterpart was a slightly built man of about five feet two inches tall with a thick, neatly-trimmed beard. His face was deeply tanned and despite his size there was an obvious toughness about him. Keeley rightly assumed that you don't get to be the RSM of an Afghan brigade without seeing some action. The two made a comical pair, Keeley with his barrel chest and upright Guardsman's posture and his tiny protégé with his neat little beard and bright green beret. They were the source of much mirth to the Grenadiers, although no one would mention this to the Sergeant Major's face. The little Afghan proved himself to be most capable and it was clear to see that Keeley's immense presence had rubbed off on him. The Afghan was soon acting the part of brigade RSM with great gusto.

Even though the cultural differences had caused some frustration on both sides, relationships were being forged. It was difficult not to like many of the easy-going Afghans, they smiled effortlessly and had a relaxed attitude. Some of the Grenadier NCOs had been less than complimentary about the ANA after their early encounters, but they had started to soften. Sergeant Clint Gillies of the Queen's Company had been a vociferous critic early on, but on several occasions when he could not be found, he was eventually located drinking chai with the Afghans. The CSM soon learned that if Gillies was not around he would be engaged in some form of Afghan social gathering. This change in attitude was striking and surprisingly mutual. Slowly an understanding and sense of trust were being forged and the Grenadiers were learning that patience was definitely a virtue where the ANA was concerned.

On their return to FOB Tombstone, the Inkerman Company was tasked at very short notice to run an NCO course for the small number of ANA soldiers who were left in Shorabak. PT was not a popular pastime for the Afghans and there was little enthusiasm for it. The Grenadiers, as always, led the way and sweated with their charges who were more accustomed to the dry heat, but there was no improvement among th Afghan soldiers. When the 90 or so students were set a fitness test, which involved running 1.5 miles as a squad and then a second circuit as a race, they seemed genuinely incredulous that the British should expect them to do this in the early morning heat. There was subsequently little enthusiasm from the Afghans who had learned how important it was to conserve energy in the baking heat. Major Elliot-Square (known as 'Box' to his men) had decided that a 'three strikes and out' policy should be instigated in order to ensure some form of quality control. Those students who consistently failed to meet the standards set, were, after a series of warnings, dismissed from the course. To the dismay of the Inkerman Company officers and NCOs, these same men reappeared the following day having been reinstated by the Afghan sergeant major, who perhaps understandably felt it necessary to stamp his own authority on the course. When it came to team competitions, though, the Afghans excelled. They were warriors at heart and hated losing face. When the Grenadiers realised this, the knowledge was used to fine-tune the training and steady progress was made.

It was soon discovered that when certificates were issued for achievement, the ANA soldiers came alive and were full of enthusiasm. This procedure was then widely adopted by all of the OMLT. Certificates were issued for all manner of skills from mechanics and driving to NCO courses. The presentation ceremonies were important affairs for the Afghans and were often carried out amid great applause and, occasionally, a burst of the Afghan national anthem. A lot of time was spent on the ranges trying to improve the standard of marksmanship and progress was made, although it was very slow. The concept of each man keeping his own weapon never quite seemed to be understood by the Afghans or, if it was, they chose to ignore it.

Now that the weapons and equipment had been accounted for and secured, the Grenadiers had a much better idea of the kandak's capability. The feeling was that once the equipment had been stored and maintained, an effective logistic system would be easier to implement. The ability of the Afghans to improvise and to circumvent the best British plans was clearly demonstrated early on. A short notice deployment order was received which required a group of ANA troops to leave Shorabak quickly. The British mentors moved rapidly to ensure that all the weapons, ammunition and equipment they would need was efficiently issued from the newly operating stores. The Grenadiers were stunned to find that the ANA men were already mounted on their vehicles festooned with ammunition and weapons including their 12.7mm DShK machine guns and RPGs. They were ready to deploy and appeared to be fairly well prepared. The British were surprised because none of the newly accounted for and neatly stored weapons had been touched. Every item of equipment seemed to have been duplicated but none of the OMLT troops could discover where the equipment had come from. ANA resourcefulness was impressive and the British were learning that nothing was as it seemed in Afghanistan.

The Grenadiers used the lessons they learned to adjust the methods they used with the Afghans in order to get the best out of everybody. It was soon found to be useless to try and organise anything in the afternoon. The ANA working day commenced at 0500 hours and by lunchtime they felt that they had generally done enough. Most Afghan soldiers would be found sleeping in the afternoons and although this frustrated the Grenadiers, they could certainly understand the logic behind the practice. On one particular afternoon some training on compound clearing was arranged in Camp Shorabak. The Inkerman Company had prepared detailed demonstrations and lessons on how to enter a compound, how to clear the buildings and in particular on the procedure necessary to safely clear a room. Forty or so ANA soldiers were expected to be present for the training and the Grenadiers were disappointed when only half that number materialised. The planned demonstrations and lessons proceeded regardless of the reduced numbers and some good results were being achieved. The sun was by now at its hottest and everyone was tired and sweating. A great deal of shouting was necessary as part of the procedure and the Afghans who had turned up for the training enthusiastically launched themselves into the building being used. Halfway through the exercise a tired-looking Afghan face appeared at a window and, through an interpreter, asked if the British and their charges could keep the noise down as his platoon were trying to sleep in the adjacent building. On the face of things this was a reasonable request until the Grenadiers realised that the sleeping platoon was in fact the missing half of the company that they were supposed to be training. This was yet another lesson learned and Afghan culture would need to be respected and changes to procedures made. It was late spring and the days were getting progressively hotter. Other useful work and training inside FOB Tombstone could easily be done in the afternoons by the British troops.

It was always expected that some unfortunate members of the OMLT would be struck down with the dreaded diarrhoea and vomiting (D&V) bug, but the early influx of anticipated casualties never appeared. This was considered to be in part due to the very high standards of cleanliness and hygiene within FOB Tombstone. Unfortunately for him, the very first D&V case was CSM Glenn Snazle of the Queen's Company. He had to endure several days of misery in the isolation room in order to reduce the risk of infection to others. A single toilet had been earmarked for D&V cases for the same reason. It wasn't long before his unsympathetic peers had christened the toilet 'The Snazle Suite' and had labelled it as such. The name was to stick for the remainder of the tour and not the slightest bit of sympathy was forthcoming from his fellow warrant officers.

While the OMLT elements were settling into Shorabak, 3 Company and the Brigade Reconnisance Force (BRF) had been carrying out their final training in the UK. Many of them were frustrated that they had been held back in England for so long, but the handover was staged over more than a month in order to allow a gradual transition in the area of operations. 3 Company arrived at the very end of March and the BRF followed a few days later. The Grenadier troops were now all in Afghanistan. Both groups attended their own packages in exactly the same way that the OMLT companies had. The exact role of 3 Company was not yet established. There was a little frustration at the uncertainty, but the following days were put to good use in training and administration. One of the problems of being formed at short notice and potentially being moved between battlegroups was that the logistic chain became confused. Equipment was always in short supply and, regardless of its nature, it was always prioritised by the various quartermasters. CQMS O'Halloran was run ragged as he tried to adequately equip the company and to establish a working administrative chain. His efforts were not helped by the uncertainty over the company's role.

After a few days it was confirmed that 3 Company would be heading to Garmsir. Major Will Mace set about planning and training his company for the task at hand and CSM Robinson, as always the driving force, provided the steadying hand that would be needed in the days ahead. Just like the OMLT troops, both

3 Company and the BRF spent the maximum available time on the weapon systems that had been in such short supply during training in the UK. Both groups experimented with explosives and learned how to use the bar mine. This device was an anti-tank mine that had been adopted for more general use. It was a rectangular block of moulded RDX/TNT explosive about 1.2 metres long, weighing 8.1kg enclosed in a plastic case and designed to explode under the heavy tracks of a tank. Two years of experience in Afghanistan had shown that the bar mine was ideal for breaching the rock-solid outer walls of the Taliban-held compounds. With some minor adjustment, a time delay could be set allowing those placing it to get under cover before the powerful mine exploded.

The BRF meanwhile completed their brief training in and around Camp Bastion, before joining the Marines whom they would be replacing in the field. An area had been identified in the desert not too far away in which they could conduct some joint training; living and operating in the desert with the people who had done the job for six months would be invaluable experience. Some intensive training had taken place in the final weeks before leaving the UK and a lot of work had been done on driving skills and weapons training. The Marines were able to provide some detailed ground briefs and good accounts of their own encounters with the enemy. Lives could well depend on these hard-earned lessons and the BRF troops soaked it all up like sponges. These few early days in the desert were most important but they passed all too soon. 12 BRF soon returned to Bastion where they said farewell to the Marines and started preparation for the 'real thing'. They would not have long to wait. Major Rob Sergeant was receiving orders from Task Force Helmand for the first BRF operation and it was due to start in just a few days' time.

6

CLOSE ENCOUNTERS

While the training and routine in Shorabak went on, the remainder of the Grenadiers were still widely dispersed. 12 Mechanised Brigade had not yet taken over formal responsibility for the area of operations. The BRF and 3 Company had only just arrived in theatre. The territorial soldiers of Somme Company with their attached Grenadiers were still getting to grips with the security and procedures required in Camp Bastion. 2 Company were settling into the routines and patrol patterns in their various locations to the north. One of these locations was Kajaki and 6 Platoon had already conducted several familiarisation patrols in the surrounding villages and fields. There had been frequent evidence of Taliban presence but they had so far escaped close contact. After only a week or so in place this changed.

An ANA platoon mentored by Second Lieutenant Howard Cordle and his Grenadiers were patrolling in the poppy fields to the south of Kajaki as part of a planned clearance operation with the Royal Anglian Company. These operations were designed to prevent the enemy from gaining a foothold in the area and to disrupt their operations so that they could not exert influence over the civilian population. They were advancing with a platoon of the Anglians to their right forward and Cordle's Afghans left forward, with a reserve platoon in the rear. They were in the area of the village of Olya, some 2km to the south of Kajaki Dam. The various compounds were quiet and many were completely deserted, although this was not unusual in the area given the intensity of the recent fighting. Cordle was forward where he could advise the ANA commander and Sergeant Ty-lee Bearder was further back where he could carry out a similar function and could also react to any difficulty the forward elements might find themselves in.

Unbeknown to the patrolling coalition troops, a deadly ambush awaited them; the Taliban lay in waiting and were now only about 100 metres or so from Cordle's lead elements. The silence was shattered by the distinctive whoosh of two RPG missiles as they shot through the air close to Cordle's position. The sound was followed by two tremendous bangs as the warheads exploded nearby. The ANA and the Grenadiers needed no encouragement to take cover and as Cordle went to ground, the walls of the building he was standing next to were raked by machine gun fire. The Taliban gunman must have been aiming at him as the ambush was sprung, but his marksmanship was thankfully no better than that of the ANA. As the first explosions rocked the air, followed by the crack of the 7.62mm rounds, Bearder quickly moved up to take stock of the situation ahead. It was clear that the lead elements of the platoon had been ambushed at quite close range from several fire positions; there were at least five or six Taliban with at least two RPGs. Bearder was able to raise Cordle on the radio and to ascertain that the young officer and the rest of the platoon were so far uninjured. The raised adrenaline and relief of both men resulted in them giggling like children as they discussed the best way to extract themselves from the ambush. Bearder now made his way forward to join his young platoon commander who had wisely taken shelter behind the mud walls of a nearby compound.

As they talked, several more loud explosions came from the area in front of them. A thin layer of black smoke could be seen drifting from there. Cordle quickly identified that he was now under mortar fire and it seemed to be accurate. Bearder, a 30-year-old veteran of Northern Ireland, Bosnia and Iraq, disagreed that mortars were the source of the explosions, he felt that more RPGs were being used and poured scorn on Cordle's assessment. As he raised himself up to better appraise the situation, a second belt of explosions impacted only 20 metres away, sending him scuttling for cover. This was confirmation that Cordle had been correct, the Taliban were using light mortars and they were pretty accurate too. Bearder was quick to admit that the young officer had been correct and he later ate a good deal of humble pie. For now, though, the fight went on.

The ANA had identified a well-concealed enemy position off to the right from where the Taliban were engaging the Anglians. They opened a heavy volume of fire on this position, although no Taliban were seen to be hit. The Royal Anglians or 'Vikings' as they called themselves, were in contact too and the rear platoon was ordered to close up to the ANA position. The British soldiers were eager to get into the fight and quickly identified that the rooftops were the best place from which to suppress the enemy. As they started to climb the steps and any other routes to the roof, the OMLT interpreter shouted a frenzied warning to the Grenadiers. The ANA soldiers were gesturing excitedly towards the Vikings now climbing on the roofs. The interpreter was quick to point out that the Taliban would be waiting for the British to occupy the rooftops; indeed, this is what they expected and hoped for. As if in reply to the warning, the volume of enemy fire increased and swept across the roofs. The British soldiers quickly dived behind the low edges of these compound roofs. It was a vital lesson that was well learned from their battle-experienced Afghan charges. The firefight, which lasted some 30 minutes or so, was brought abruptly to a close when coalition jets delivered several high explosive bombs onto the area of the enemy positions. It was likely that most of the Taliban fighters had withdrawn already. They knew only too well the destructive power of the allied aircraft.

On this day there were no friendly forces casualties and it was impossible to confirm the number of Taliban killed or injured. It was likely, however, that a handful had been killed in the air strikes. The coalition troops patrolled back into their base at Kajaki. They had confirmed the forward line of enemy troops and now knew their capability. Later there would be a debrief and many valuable lessons would be drawn from this first contact. There had most definitely been an element of luck involved and those who had been under fire were well aware of this fact. When the Grenadiers reached their OMLT building, they immediately started their post-action administration. Weapons and equipment were checked and ammunition was replenished, magazines refilled and 6 Platoon was once again prepared for short notice action. The ANA by contrast had largely disappeared; they immediately found shade and sparked up their cigarettes. They were much more experienced in combat than the British and were more relaxed in the aftermath. Sergeant Bearder made a note to encourage the ANA platoon sergeant to check his men and their equipment after action, but realised the Afghan had seen a lot of fighting in his time.

The Grenadiers were still buzzing from their first action, but as the adrenaline wore off they suddenly felt very tired, thirsty and in need of some sleep. Many lessons had been learned and these were discussed in an 'after action review' which was held by Cordle and Bearder. It was noted that the ANA observation was outstanding and that they had located the enemy before anyone else. Their warning about using the roof had been timely and they obviously had plenty of operational experience to pass on to their 'mentors'. As far as most of 6 Platoon were concerned, the biggest plus had been the performance of their new platoon commander. Cordle was straight 'out of the factory' but he had done exceptionally well and the Grenadiers were delighted that they had a very good young officer leading them in these dangerous circumstances. The platoon would soon learn that Cordle and Bearder were a very strong and competent team.

That evening Cordle received orders for another operation the following day. The OMLT and ANA would be operating in support of the Anglian Company and a large area was to be cleared. They would be moving out before it got light and everyone knew that the surrounding area was full of Taliban. Later, orders were given and preparations were made for the operation. When everything was done, the Grenadiers and their Afghan colleagues settled down to get some sleep.

In addition to the Grenadier OMLT troops, C Company of 1 Royal Anglian were also in Kajaki. One of the sections in C Company was made up of detached Grenadiers who were in 10 Platoon. They too had been involved in the recent engagements with the Taliban. They had arrived in Kajaki with the remainder of the company from the eastern town of Now Zad in early April. At first things in Now Zad had been quiet. All that changed on 13 April when the company took its first fatal casualty in a fierce firefight. This was 12 Brigade's first fatality and it came only two days after the brigade formally assumed responsibility for Helmand Province.

The men of 10 Platoon's 3 Section, commanded by Lance Corporal Thompson, had seen plenty of fighting. 'Tommo' had come from the Grenadier's Corps of Drums and was now responsible for eight Grenadiers who were a long way from their own battalion. They had learned quickly from their experiences in Now Zad and in Kajaki. They were becoming familiar with the ground and the way that the Taliban operated. It was necessary to mount frequent clearance operations around the Kajaki base in order to prevent the Taliban from gaining a foothold close to the dam from where they could operate. The infantry soldiers from C Company and from the Grenadier OMLT found themselves in the small villages almost daily. While clearing the myriad compounds it was quite possible to meet the enemy at very close range. Usually the Taliban would try to mount a quick ambush and would withdraw, but sometimes the opposing forces stumbled across each other and close quarter fighting would result. On one occasion 10 Platoon stumbled on a group of Taliban who had been washing in a stream. They were literally just around the next bend. A fierce firefight ensued in the close country and both sides were incredibly fortunate to escape without serious injury. There were no casualties on this occasion, but during another clearance operation 10 Platoon were not so lucky.

C Company had been clearing a series of compounds to the south of Kajaki and 10 Platoon gained entry into one of their objectives, a collection of low buildings surrounded by the usual high mud walls. The compound was deserted and there was no sign of the enemy, although movement had been observed in the area the previous night. 3 Section cleared the buildings in sequence, ensuring that every room was systematically searched for signs of the Taliban. This was a straightforward procedure and was completed quickly. The next phase was to 'bounce' on to the next compound and to repeat the process. 3 Section gathered along one of the hard mud walls close to a gate. They waited for their turn to move. The other sections started to exit through a different gap in the wall and as they did so the sound of automatic gunfire shattered the silence. On the radio Thompson could hear that 10 Platoon had taken a casualty. The platoon radio operator had been hit. The enemy had opened a withering fire on the British troops and those who could returned the fire with interest. 3 Section being further to the right were better placed to move forward and to try and get around the flank of their attackers. Thompson received instructions to do just this. He peered around the gateway and surveyed the ground ahead. The next compound was some 30 metres away across open ground. They would have to cross at a sprint in order to reach the relative safety of the compound. Thirty metres is a long way to run when you are under fire and Tommo looked for cover. Almost in the middle of the open ground, there stood a small building. It was just a couple of metres square with half a roof on it. He had no idea what it had been used for but it would provide some cover from fire.

Tommo shouted his orders and the Grenadiers of 3 Section were soon sprinting hell for leather for the safety of the building's mud walls. First to arrive were Guardsman Alex Harrison and Guardsman Chris Bangham. The two 19-year-olds were conscious that the compounds ahead of them, now only 15 metres away, might be concealing Taliban fighters. Harrison cautiously peered around the left edge of the small building and then moved quickly towards the far corner, all the time looking for enemy in the compounds ahead. Bangham prepared to follow. As Harrison passed the small open doorway to the tiny building at his right shoulder, gunshots rang out and he collapsed to the floor. It was obvious to Bangham that Harrison had been hit in the head. It was unclear where the shots had been fired from, but as he moved forward to assist his colleague, a second burst passed just in front of his chest. Bangham felt the force of the 7.62mm bullets as they passed inches from him and realised that they had been fired from inside the small building. He recoiled, shocked that the Taliban fighter who had just shot his friend was now only a couple of feet from him on the other side of the wall. Bangham and the other Grenadiers prepared high explosive grenades to deal with the fighters within and as they did so Harrison called out, indicating that he was not only alive but conscious. Bangham and two others from 3 Section lobbed the L109 HE grenades over the wall. Seconds later they detonated with deep thuds within the tight confines of the tiny structure. Black smoke and dust rose from inside and poured out of the tiny doorway that minutes earlier had concealed the enemy. A brief groan was heard from within after the first detonation but nothing further after the second and third. On the other side of the building, Tommo had noticed a previously unseen trench that linked the small building to the next set of compounds. This had probably been some sort of sentry post. Its occupant would not be performing any duties there in future.

The main priority now was to extract Harrison from his exposed position on the other side of the small building. Fire was being directed at them from elsewhere and tracer bullets were tearing up the ground around the gravely injured man. Supporting fire from the other sections in 10 Platoon added to the confusion. Bangham and the others called to Harrison and, to their amazement, the seriously wounded man quickly replied. They asked if he could get back to the compound and he replied that he could. 3 Section put down covering fire onto the likely enemy positions to their front and, as they did so, they witnessed the incredible sight of their wounded comrade rising to his feet and running back towards their original position in the cleared compound as tracer rounds passed inches from him. Harrison disappeared through the gap and Tommo extracted his section back across the open ground. Each man sprinted across the exposed area and their section commander was relieved to see them all make it safely back. Harrison was now seated, his back against the thick mud wall. Other soldiers were attempting to treat his obviously serious head injury. The enemy fighter had shot him from very close range as he passed the door . The bullet had skimmed his right ear and entered his head just behind his temple, exiting through his right eye socket, blinding him in that eye at the same time. For those looking on it was difficult to understand how he was alive, much less still conscious. His head had already swollen badly and there was a lot of blood. Harrison, however, seemed oblivious to his life-threatening injuries and he continued to talk to those attending him. As Bangham watched, he saw his friend draw a map in the sand using a spent cartridge. Harrison had been the lead man in the section and had got further forward than anyone else. He had spotted the enemy positions which had so far eluded the others and he now passed on this information with complete disregard for the seriousness of his own situation.

Guardsman Alex Harrison's actions were in the highest traditions of the Grenadier Guards. His selfless commitment to his comrades was all the more impressive for a man of only 19 years of age. 10 Platoon were able to extract, but the company took four casualties that day. Harrison was carried on the back of his platoon sergeant, Steve Armon of the Royal Anglians, before he could be safely lifted out to Camp Bastion. When he arrived at the field hospital his injuries were so serious that there were concerns that he might not make it. The surgeons and medical staff worked very hard to stabilise him, but his condition was critical. The doctors and nurses worked around the clock to save Harrison's life and although he remained critically injured, his condition was stabilised. It was then possible to fly him back to the UK, where he could be admitted to one of the specialist hospitals which were better able to deal with his condition. Harrison was still in a critical condition and every Grenadier in Afghanistan prayed that he would pull through.

7

RECONNAISSANCE IN FORCE

Like the other Grenadiers, 12 BRF were now on their own. The last of the Marines had gone and any training still to be done would take place on the job. Major Rob Sergeant had received orders for a reconnaissance task in the area between Gereshk and Sangin. This district had long been a Taliban stronghold and they had been able to intimidate the local population from the villages in and around the Green Zone and had launched repeated attacks against coalition forces. 12 Brigade were planning a major operation in this area which was aimed at clearing the enemy out of the Upper Gereshk valley for good. Unfortunately, the intelligence on the exact Taliban positions was incomplete. Before a major operation could be mounted, more information on the enemy strengths and dispositions would be required and this is where 12 BRF came in.

The specialist troops were to establish routes suitable for friendly forces vehicles and crossing points over the various canals and waterways leading to and from the Helmand River. They would need to ascertain the pattern of life in the villages and establish where the Taliban strong points were. During their training they had done a lot of covert surveillance and close reconnaissance work, but the Marines had warned them that the Taliban were hard to find by this means. They would have to be prepared to fight for information. The plan was for the BRF to probe the likely enemy strongholds in a number of villages. This preliminary reconnaissance was a part of what the Army calls 'shaping operations'. They would move from village to village over a period of several days, testing the enemy response. Other assets would be used to observe the Taliban reactions and this intelligence would inform those responsible for formulating the detailed plan. If all went well, 12 BRF would identify targets for the forthcoming operation, which was now named Operation Silicon. The reconnaissance troops also hoped to identify potential enemy reinforcement routes and command structures. If they were able to stir things up sufficiently, the BRF would also be able to deceive the enemy as to the brigade's intentions.

When the platoons returned from their brief excursion into the desert where they had trained, they set about preparing for their first real operation. All of the usual battle preparation took place: ammunition storage, weapons preparation, vehicle maintenance and a whole host of other tasks. The BRF received their orders from Sergeant and only a few days after coming in from the harsh desert, they were heading out again. This excursion would be a lot longer, but this time they would also be actively looking for a fight with the Taliban. As the convoy of WMIKs and Pinzgauer six-wheeled vehicles drove out into the desert from Bastion, the heavily laden vehicles generated huge dust clouds as they meandered away from civilisation. Everyone on board was soon coated in thick dust and the well-maintained and oiled machine guns turned brown as the desert soil crept into every screw and depression. Some people commented on how the sight resembled pictures of the famous Long Range Desert Group of Second World War fame. On the face of things there really was very little difference: dust-covered jeeps and men with machine guns, struggling against the blistering desert sun. What was not immediately visible was the impressive firepower and additional specialist capabilities owned by the BRF. Quite apart from the obvious sight of the .50 calibre heavy machine guns and GMGs, there were snipers and mortar men in the platoons. There were two 81mm mortar barrels capable of dropping high explosives onto the enemy from several kilometres. There were also men from the Grenadiers' Anti-Tank Platoon who would operate the very potent Javelin missile system. The Javelin was a fast, hi-tech set of equipment which when fired would deliver a very powerful missile with incredible accuracy.

There were other specialists whose jobs were equally important. Men from the Royal Signals would be responsible for signals intelligence and were known as the Light Electronic Warfare Team (LEWT). The BRF also had the capability to launch its own small unmanned aerial vehicles (UAVs). These were in effect small pilotless aircraft fitted with cameras which were capable of flying over and filming an enemy position. A 'real-time' video would be relayed to a control station mounted on one of the vehicles. Commanders could then study the terrain and even enemy positions from the relative safety of a command vehicle. Controlling the small aircraft and cameras was a skilled job which fell to men from the Royal Artillery. Other gunners in the BRF were trained to direct air strikes and artillery, but every man was capable of directing the 105mm gunfire onto the Taliban. The mechanics from the REME were also vital to the mission. All of the WMIKs had taken a pounding in the harsh terrain but the platoons could not afford to lose broken-down vehicles in the desert. Meanwhile, the medics, everyone knew, would be the most important people in the team. If someone was hit, it could take some time before the Medical Emergency Response Team (MERT) arrived by helicopter from Bastion. The medics would be responsible for keeping the casualties alive until they could be lifted out to the field hospital back at Bastion. The level of care there was superb if only men could be kept alive for that critical time after wounding.

The first task for the BRF was at a village to the north of Gereshk. It was believed that the enemy were very active in the area but little was known about how they were organised or how strongly the area was defended, if at all. It was incredibly difficult to tell the difference between a Taliban fighter and one of the many local civilians. If they concealed their weapons or discarded them it was impossible to know who was who. There were no uniforms and the Taliban had no qualms about using civilians for cover. The inhabitants of the villages were at the mercy of these ruthless fighters, many of whom were foreign. It had been decided that the best option would be to approach the villages at night. The BRF would make a 'demonstration'; in other words, they would let the Taliban know they were there by pushing into the Green Zone around the village. It was expected that the enemy would respond by attacking the British troops. The BRF would then draw them out where they could be assessed by both ground and air assets. In this way much intelligence could be gathered about how many fighters there were, what weaponry they had and from where they were likely to be reinforced.

Major Rob Sergeant had decided upon a plan for a night approach. One of the two platoons would be dismounted and would move towards the village on foot. The second platoon would remain in its vehicles and would man the heavy weapons that would be so important if the Taliban decided to play. The mounted platoon was split into two halves, one on each side of the dismounted platoon as it moved towards the village. Daytime reconnaissance and detailed studies of the map had revealed a wadi which ran west to east from the village. In places it was deep with very steep sides and would provide a good covered approach for the dismounted platoon. They would be safe from ambush because their flanks were protected by the mobile troops. The planned time for operations to commence – broadly known as H-hour – was set for 0330 hours and as the troops moved off they noted that it was an exceptionally dark night. Even their sophisti-cated night vision aids were degraded by the low light levels. The WMIK drivers moved slowly as they struggled to see the ground ahead through their infra-red helmet-mounted monocles. CSM

Ian Farrell was at the rear with company HQ when he heard disastrous news over the radio. One of the WMIKs moving to the north of the wadi had rolled. The restricted view through the night vision device had caused the driver to misjudge the ground ahead and the Land Rover had turned over on the uneven ground. Fortunately there were no casualties, other than the vehicle which had been quite badly damaged and could not be used again that night. Farrell moved up to the stricken vehicle in his Pinzgauer and with the aid of the REME fitter, was able to turn the vehicle back onto its wheels. This whole operation had to be conducted in complete darkness and as quietly as possible. While they were still some distance from the village, they did not want to alert the enemy to their presence until they were ready.

H-hour was delayed by the best part of an hour and a few rearrangements were made before the platoons got themselves in position. The dismounted platoon moved eastwards on foot up the wadi with the remaining WMIKs either side of them, slowly rolling along so as to reduce the engine noise. Sergeant and his company HQ were a short distance to the rear and the CSM was behind them. Farrell would be responsible for ammunition resupply and casualty evacuation in his Pinzgauer if the need arose. The complete lack of ambient light was causing problems and the dismounted platoon initially became disorientated in the area of a brick factory. After the initial confusion they were able to move on steadily towards their objective, conscious that on foot they were very vulnerable. Lieutenant Andrew Seddon, the platoon commander, hoped that the mounted troops were in good positions.

As the mounted platoon moved towards their overwatch positions on either side of the wadi, the troops on the north side suddenly picked up movement through their night vision devices. There had been no other sign of enemy activity and until now things had seemed calm. Four Taliban were seen as they left the village mosque, they rapidly produced an RPG, loaded and fired at the group of vehicles to the north of the wadi. The rocket could be seen and heard as it streaked through the air.

Things now happened quickly. Additional explosions were heard to the north and seconds later there were three detonations near to company HQ and the southern mounted platoon. At first there was confusion and then the realisation that these had all come from RPGs. The northern overwatch position had been engaged by at least three of the deadly rocket launchers from a compound about 250 metres to their north. The rockets had overshot and had come down very close to Captain Andy Mac, the second in command, and his vehicles. The enemy were obviously firing blind because of the intense darkness, but had almost got lucky. Fire was returned at the northern compound but, shortly after, enemy positions on the western side of the village opened fire. They were joined by more Taliban who had taken up positions to the south. Seddon's dismounted troops spotted a group of enemy fighters as they ran across in front of them. They were heading to the north to assist in the fight, but were rapidly cut down by the reconnaissance soldiers. The BRF was now under fire from three sides. If the intention was to stir things up, they had been successful. There was a great deal of excited chatter on the radio nets. For most of the soldiers this was their first real action and there was uncertainty and apprehension, but this was short-lived. Sergeant was soon heard on the radio directing the battle and this was most reassuring. Training kicked in and as the BRF seized control of the battle, the uncertainty disappeared. The enemy fire became progressively stronger as more Taliban joined the fight and the position to the north seemed to have been reinforced. The recce troops fired back, relying heavily on the .50 calibre Brownings and the GPMGs mounted on the vehicles commanded by Lieutenant Mike Holgate. Orange tracer arced through the darkness onto the Taliban defensive areas and bright flashes briefly illuminated the immediate area as RPG rockets detonated. Both sides hammered away at each other, the Taliban uncertain of British intentions and the British content to allow the enemy to expose their positions to the many deployed surveillance assets.

The northern compound was causing great concern as volleys of RPG rockets continued to explode around the northern flank of the mobile troops. This position would have to be neutralised before one of the rockets hit home. The Fire Support Team (FST) had been busy and at their request air support arrived on station. The FST directed the US A10 'Warthog' pilot onto the target compound. To the British troops the target area was obvious. Tracer ammunition was coming from and being directed at the compound. Things were not so simple for the US pilot who was several thousand feet above the target, as tracer could be seen ricocheting around the entire area. There were many compounds and most of them had heat signatures that could be plainly seen through the thermal imaging devices fitted to the aircraft. Unfortunately the technology did not distinguish the good guys from the bad. The American officer had to be positive that he was not about to release his bombs onto friendly forces or onto a house full of innocent civilians. The twin-engine aircraft made several passes but the pilot was unable to positively identify the target.

As the jet circled above them the British troops continued to pour fire into the compound. The controllers on the ground now noticed that one of the Taliban bodies had been set alight by a tracer round. This could be plainly seen through the night sights and thermal imagers. The location of the burning body was passed to the US pilot and after a short delay he confirmed that he could see the body and could now identify the target compound. He circled the area once more and prepared to make his bomb run. The troops below hugged the ground in anticipation of the deadly detonation. Jet engines roared and the high-pitched whine of the motors signalled that the aircraft was accelerating away at 6,000 feet per minute. The pilot indicated that his bomb had been dropped and everyone waited for the huge explosion. Nothing happened and the Taliban continued to pour fire in the direction of the BRF troops. The pilot confirmed that his bomb had been dropped and that it must have been a 'blind', the military term used to describe a dud. This was very frustrating but the enemy in the compound still had to be silenced. Eventually the Grenadier anti-tank men fired two Javelin missiles. There was a loud whoosh as each missile was launched and the tracer flare in the rear of the projectiles could be seen streaking towards the target. The little dots of light suddenly shot upwards and then dived down at incredible speed before detonating. Sparks and flame could be seen briefly in the darkness above the mud-walled buildings. The firing from the northern compound stopped immediately. If anyone was left alive they had obviously decided that the compound was an unhealthy place to be.

By now the fire from the other two positions had stopped too. The enemy had either been killed or had withdrawn. So far the mission had been a success. The Taliban had done exactly what the brigade intelligence personnel wanted. Much had been learned about the enemy in this area who had been mauled by the BRF firepower with no friendly forces casualties. It was now time to get out of the area before first light when they would become more vulnerable to Taliban fire. To further confuse the enemy, a number of bar mines were blown, each of these detonated with a tremendous roar which could be heard for miles. The British used the brief window of enemy uncertainty to break contact. The dismounted troops rapidly withdrew and after a brief reorganisation the whole BRF mounted up and drove away from the area. There was great relief and excitement at having come off best in their first encounter with the Taliban.

Their route took them south through Gereshk and eventually the familiar antennas and flag poles of Camp Bastion came into view. It was daylight now and the strain on the faces of the tired soldiers was visible. Everyone was pleased to be back in the secure area with its little luxuries but no one could relax yet, there was much to be done. The BRF command group went to brief the commander of Battlegroup North on what they had discovered about the enemy. This battlegroup would be moving towards the area during the forthcoming Operation Silicon. While the officers briefed the commanding officer of the Royal Anglians, the remainder of the BRF carried out their essential administration. Weapons were cleaned and ammunition replenished. Fresh stores were again stowed on the battered WMIKs, many of which now needed urgent attention from the REME. The troops swapped stories of the recent events. It had been a long night and the contact had lasted for several hours. Some had been surprised by the weight of fire the Taliban had brought onto them, but they had proved themselves to be more than a match for the enemy. The chatter went on for several hours. The commanders sat and analysed the events. The things that had gone well were quickly identified and lauded and the efforts that had not gone to plan were also thrashed out. It was important to learn the lessons before the next encounter with the enemy. Behind the euphoria they all knew that they had been lucky: the element of surprise had been with them and their weapons systems were superior to those of the enemy. It would still have been very easy for someone to have been killed in the firefight. This was in the back of some people's minds and they knew very well that this was only day one of the operation and there were several other targets to be probed. But by now everyone was exhausted and the administration completed, so nearly everyone stretched out to get some much-needed sleep.

A day or so later, orders were received that the BRF was to proceed as planned against the other enemy areas and they once again headed out into the desert. The next target village was called

Pasab and was a collection of compounds virtually indistinguishable in character from the previous target area. It too was surrounded by poppy fields and irrigation ditches. There was a vital crossing over the river near the village and the plan involved the temporary seizure of the bridge. Seddon was to occupy some of the nearby compounds on the far bank and was to defend the area against the anticipated Taliban attack. Intelligence suggested that the area was important to the enemy and that they would mount a rapid attack against any coalition forces that looked as if they might remain in the area. Not much was known about the village, so the plan was to lay up in the desert and to mount surveillance on the area to establish the pattern of life in the small hamlet. The BRF would need to find the crossing points over the various watercourses because they posed major obstacles. If the Taliban were visible, their strong points were to be identified. Observation posts were set up and the British soldiers sweated in the afternoon sun as they watched the farmers working in the poppy fields. No Taliban were seen but there was much activity in the village as people went about their business.

The following morning the plan was to be put into action. It was decided that this time the BRF would approach the village in daylight and H-hour was set for 0700 hours. Some people managed to snatch some sleep but they were aware that the enemy was close. It was decided to send up the 'Desert Hawk', the UAV, which would circle over the area to give an up-to-date reconnaissance picture. It was a wise decision and the little aircraft was able to relay imagery which showed clearly that the bridge had been blown. It was now necessary to change the plan and Seddon was ordered to move into the village instead of over the bridge. The vehicles were secured out of sight in dead ground and just as in the previous operation the dismounted platoon approached the village. They were covered by their colleagues in their WMIKs, who scanned the edges of the settlement for signs of the Taliban.

© Alexander Allan

TOP: Major Marcus Elliot-Square (officer commanding the Inkerman Company) with Colonel Rassoul, the commander of 3rd Kandak.

ABOVE: Lieutenant Colonel Carew Hatherley, commanding officer of 1st Battalion Grenadier Guards. Hatherley's battlegroup was widely dispersed over Helmand. In addition he acted as mentor and advisor to the Commander of 3/205 Brigade of the Afghan National Army.

RIGHT: Regimental Sergeant Major 'Stumpy' Keeley with his Afghan counterpart.

© Alexander Allan

ABOVE: Royal Engineers work to construct a new patrol base on the banks of the Helmand River, north of Gereshk. This area was to be fiercely contested by the Taliban.

BELOW: The Inkerman Company getting to grips with the Afghan terrain early in their tour.

ABOVE: OMLT mentors on the range with their charges from the Afghan National Army. Camp Shorabak, spring 2007.

BELOW: Soldiers from 2nd Kandak 3/205 Brigade of the ANA – war weary but fierce and courageous when well led.

ABOVE: The charismatic Colonel Rassoul addresses
the local population near Babaji, spring 2007.

ABOVE: Afghan troops in
the field.

LEFT: The interior of an
OMLT WMIK, showing the
7.62mm general purpose
machine gun mounted
in front of the
commander's seat.

ABOVE: An ANA patrol moving
through an Afghan village,
early summer 2007.

BELOW: A Grenadier prepares to
move on a night time fighting
patrol, Garmsir, summer 2007.

© Alexander Allan

LEFT: Guardsmen Harrison and Bangham, Grenadier Guards who were attached to the Royal Anglians at Kajaki. Harrison was shot in the head and severely wounded early in the tour.

BELOW: The workhorse of Helmand – a Chinook helicopter loads up at Sangin.

ABOVE: The Eastern Checkpoint, Garmsir. Guardsman Simon Davison was killed fighting from the roof in May 2007.

BELOW: An ANA Ranger destroyed in an IED strike. This was an all too familiar event in the summer of 2007, and many Afghan soldiers lost their lives in this way.

ABOVE: Soldiers from the Inkerman Company and the ANA wait during Operation Lastay Kulang.

BELOW: A rescue attempt on an Afghan vehicle that was driven into the Helmand River from a forward operating base. Members of the Queen's Company look on in amazement.

It was still quite early in the morning but the fields were full of workers. The civilians were busy harvesting the notorious Helmand poppy and only their top halves were visible in the red and pink flowered fields. The advancing British troops were now about 600 metres away and were suddenly spotted by the farmers and their labourers. The Afghans' reaction was rapid. Immediately they started back towards the village, leaving the harvested poppies where they were. Tractors, cars and even minibuses appeared from the village and people crowded into them hastily. Finally the vehicles and the civilians all returned to the village, leaving nothing but a thinning cloud of dust and the now deserted poppies blowing in the morning wind.

The BRF soldiers were alarmed at this rapid withdrawal but they had so far not been engaged by the enemy. Perhaps the farmers were afraid of the British or maybe they thought that the BRF were a part of a poppy eradication force. Either way this was not good news and the last couple of hundred metres to the edge of the village was very tense. The platoon interpreter was able to relay local intelligence to Seddon. The enemy were apparently telling the civilians to get out of the village as they were going to attack. Seddon was alarmed to be told that the enemy was relaying the exact positions of the slow-moving British force. The troops were relieved when they reached the solid walls of the hamlet, at least there would be some cover there. They spread themselves out overlooking three small foot bridges and observed the various compounds for signs of the enemy. Seddon now looked for the village elders; it was important to reassure the civilians and to try and gain some information on the whereabouts of the enemy. There were few civilians around but he did manage to locate an old man who complained bitterly that his wife had run away without making his breakfast and he was starving.

Outside the village, Sergeant was greatly relieved that his troops had not been caught in the open by the enemy. Seddon had indicated that so far there was no sign of the enemy in the settlement. The troops in overwatch had seen no Taliban movement, just the village routine. A few women in their black, full-length burqas were seen slowly walking about the streets and things seemed to be very quiet. Sergeant decided to move into the village in his vehicle – this way he could get a better indication of the situation and would be able to speak to the locals himself. Before long, Sergeant was standing next to Seddon. Sergeant continued the conversation with the old man and Seddon started to move forwards. The elderly bearded figure conversed with them through the BRF interpreter. The most burning and obvious question was, 'Where are the Taliban?' The old man innocently shrugged his shoulders, gestured with his open hands and told the interpreter that there were no Taliban in the area. No sooner had the old man spoken than all hell broke loose. AK-47 bullets thudded into the ground and buried themselves in the mud of the compound walls. They cracked overhead and threw up dust clouds in the primitive streets. A burst of machine gun fire impacted on the wall just above Seddon, sending clouds of dust into the air. Sergeant and the other troops quickly dived for cover as an RPG rocket exploded close by. The British troops were now returning fire at their attackers who were very close and well positioned in several different firing points. The troops in the overwatch role had been engaged too and as they searched for the enemy positions they saw burqas being discarded as the newly revealed Taliban fighters took up arms against them. The fire was now intense as the British fired everything they had at the enemy in an effort to win.

One of the WMIKs, commanded by Lance Sergeant Stott, was engaged by an RPG which fortunately missed and detonated with a terrific bang close by. Guardsman McGhee, who was standing in the cupola of the WMIK, immediately traversed the .50 cal and commenced firing. As he did so a second rocket was fired, this time it ricocheted off the ground and exploded on the back of the

WMIK. The explosion was terrific and the driver, Lance Corporal Rob Pointin reacted by slamming the vehicle into reverse to get out of the killing area. The WMIK crashed into a depression with a thump and as he looked back to see what was happening he realised that McGhee was down. McGhee looked to be unconscious and there was a great deal of blood. The RPG rocket had detonated against the heavy .50 calibre machine gun, spraying shrapnel through the air. McGhee had been hit by the blast and by the shrapnel; he looked as though he was in a bad way. Pointin now forced the vehicle into gear and raced towards Farrell's two Pinzgauers. Stott, Pointin and Farrell hastily removed the badly wounded man from the vehicle and Lance Corporal Melville, the medic, went to work. McGhee had serious shrapnel injuries to his head and chest. He also had a very serious injury to his lower arm and was losing a lot of blood. Melville and Farrell applied first field dressings to stop the flow of blood and then placed a combat tourniquet above the arm injury, which was bleeding profusely. MaGhee's breathing was irregular and Farrell realised that an urgent casualty evacuation was needed. A request had been sent as soon as Stott and Pointin realised that they had a serious casualty and a Chinook was already lifting off from Camp Bastion.

By now the whole of the BRF was in contact and heavy fire was being exchanged. The enemy were estimated as being at least 40 strong at this point. The emergency helicopter landing site was a pre-designated area, the grid reference of which had been passed to the incoming Chinook. It was now imperative that they move McGhee to the point where he could be extracted immediately. The casualty was in the CSM's Pinzgauer with Melville. The vehicle bounced along on the uneven ground and Stott's WMIK and one other tagged along as an escort in case they ran into trouble. The moving vehicles now attracted more enemy fire which followed them as they moved off. More RPG rockets exploded intermittently, throwing up large clouds of dust. Farrell and his little party drove on through the hail of fire, aware that time was not on their side. When they reached the designated landing site the small arms fire had greatly reduced. The Chinook with the medical team on board was only minutes out, but now the CSM's little party faced a new hazard. A loud 'crump' nearby sent everyone diving for cover. The explosion was followed by several others, 'crump', 'crump', as enemy mortar rounds impacted around them. Dust hung in the air between the explosions, which appeared to be getting closer.

There was no chance of landing the Chinook here as the Taliban obviously had the area under observation and were bringing accurate mortar fire to bear. There was nothing for it but to move the location of the landing site. Farrell led the way out of this killing zone and passed the new grid reference to the approaching helicopters. As the wounded McGhee was unloaded from the Pinzgauer at the new site, the familiar 'crump' of closing mortar bombs was heard again. The little team now noticed that a pickup truck and a Hiace van had moved their location as though tracking the progress of the CSM's party. A view through binoculars soon confirmed that enemy fighters were firing from the vehicles and a man speaking into a radio was directing the mortar fire. Pointin raised his rifle and after positively identifying the enemy fighters he opened fire. For an experienced shooter a shot at 300 metres was straightforward and the enemy mortar fire controller was quickly neutralised. The little party put down a heavy rate of fire from their GPMGs, which were particularly effective at this range. Farrell fired high-explosive bombs from his 51mm mortar and then used the under-slung grenade launcher on his rifle to harass the enemy party. The Taliban mortar fire was now less effective because it was firing 'blind' and Farrell once again moved the party to a new helicopter landing site. Melville had been working on McGee's injuries and the wounded soldier was now regaining consciousness, although he still had difficulty breathing.

The Chinook soon appeared, flying low over the desert. It was accompanied by two Apache attack helicopters, known on the radio as 'Uglies'. The remainder of the BRF were still engaged in heavy fighting in and around the village and to hear that Uglies had arrived on the position was a great boost to morale. These helicopters had an awesome array of surveillance assets and firepower which would give the ground troops a huge advantage over their attackers. The twin-rotor Chinook made a rapid approach to the helicopter landing site and as it set down, its nose rising slightly, a huge brown dust cloud consumed the surrounding area. McGhee was, surprisingly, able to walk with assistance to the helicopter where he was helped up the short ramp by a medic. No sooner was he aboard than the big green machine lifted into the air and turned away towards Camp Bastion. On board were a team of medics including doctors who got straight to work on McGhee's serious injuries. It was a short distance to Bastion and about 15 minutes later the wounded man was being admitted to the field hospital.

After the successful casualty extraction, the CSM's little group now had a further important task to carry out. The fighting in the village, which had been going on for quite some time, was intense and ammunition was running low. Farrell and his crew needed to get the Pinzgauer forwards and into the village. Sergeant Tindle, who was driving the CSM's vehicle, raced as quickly as he could because the vehicles were once again under fire. As they approached the village they presented a tempting target to the Taliban and the fire increased. Occasionally, bullets could be felt striking the little vehicle and the crack of high velocity rounds could be heard over the roar of the engine. Before long they were inside the village and the CSM ordered his group to stop and to dismount. This area was the safest he could find for now, but they were still under withering fire from several enemy positions. Bullets passed close by as the little group hastily unloaded the reserve ammunition from the rear of the Pinzgauer. The situation was so dangerous that it was necessary to complete the task on hands and knees. Moving the heavy boxes of ammunition in the heat of the day, wearing body armour and a helmet, was exhausting. The process was made more difficult by the fact that everyone had to stay as low as possible. Farrell had made contact with Colour Sergeant Frith who urgently needed the ammo for his hard-pressed platoon. The two had arranged to meet halfway and Farrell, Tindle, Melville and Lance Corporal Hall now headed off, heavily burdened by the weight of the desperately-needed ammo. Shortly afterwards, they were met by Frith. Farrell was amazed to see that the NCO had acquired a wheelbarrow in which to transport some of the ammunition. Once the ammo was transferred, the CSM's party were treated to the sight of Frith and his men rushing off as fast as they could with Frith pushing the barrow full of ammo like a sort of demented gardener.

Next, Farrell moved to Sergeant's position. The officer had made a plan for the extraction of the BRF. There was nothing further to be gained in the village; they had established that the area was heavily defended and the focus now was to get out before they took more casualties, and this was agreed by the brigade staff who were keenly monitoring the situation by radio. Although Seddon's platoon was well positioned to push further into the village, enough intelligence had been gathered and this would be an unnecessary risk. Sergeant briefed the CSM on the plan he had devised. Farrell was to return to the 'Zulu muster', the area where the spare vehicles had been parked. He was to collect the remaining Pinzgauers and lead them back to the edge of the village. Once they were in position, the dismounted troops would conduct a fighting withdrawal, they would mount the vehicles and the BRF would make its escape supported by the Uglies above.

The CSM's party were soon heading away from the village, still under fire. They reached the Zulu muster and returned in accordance with Sergeant's plan. Farrell dismounted the available troops and set up a firing line in order to support those who were to withdraw. By now the firefight had been won. The enemy fire had decreased markedly although there were still occasional bursts. Sergeant and his crew made a dash for their vehicle under cover of the newly arrived fire support and quickly made their escape from the killing zone. The scene was still rather chaotic although the initiative remained firmly with the BRF. Automatic fire from both sides continued to tear lumps out of the ancient mud walls all around. The dismounted troops now began their withdrawal in a coordinated fashion. Small groups and even pairs took it in turns to fire and then to run back towards the relative safety of the CSM's group. Thankfully no one was hit during the extraction. Many of the enemy positions had now been seen close up. They were well prepared and sandbagged. The British troops tossed grenades into the bunkers as they withdrew. After a quick head count to ensure that no one was left behind, the troops mounted the little convoy of Pinzgauers and headed away. Fire was exchanged by both sides until the BRF was out of range. The Apaches, which were still overhead, had been restricted by the presence of the British troops in the village. They could now be heard firing short bursts from their devastating M230 30mm chain guns. As the BRF distanced themselves from the Taliban once again, the Uglies could still be seen hovering over the Taliban position like angry bees. They continued to fire accurate and devastating bursts into the enemy positions whenever they identified them.

The convoy of battered WMIKs and Pinzgauers moved away from the contest which was continuing in the village. They once again moved to a desert leaguer, this time just a few kilometres away. Everyone knew the routine by now. The first priority was to secure the immediate area to ensure that they were not surprised. Sentries were posted and the troops began the checks of their equipment. It was hard to ignore the bullet holes that had appeared in some of the vehicles and the CSM's Pinzgauer seemed to have taken several hits. Everyone was now low on ammunition. The contact had lasted for several hours and a replenishment or 'replen' was urgently needed. A request was sent for extra ammunition from Bastion and in the meantime everyone set about cleaning the weapons that had proved so valuable that morning. As they went about their administration the BRF troops watched the Uglies circling over the village. The optics mounted on the helicopters allowed them to identify the Taliban positions from a distance and to use the lethal array of weaponry on board to destroy the enemy. The Apache crews were able to identify 15 to 20 Taliban dead in the village. Although this had been a tough fight and everyone remained concerned about the injured Guardsman, McGhee, there was satisfaction that they had once again proved themselves to be better than the enemy. Quite apart from 'smashing' the enemy in the village, the BRF had really stirred things up and the various surveillance assets had gleaned valuable information about the Taliban in the area. During the afternoon, a Chinook put down near the leaguer and the troops offloaded a huge amount of ammunition and other supplies which were needed to continue the mission. Farrell distributed the haul around the various platoons and stored the reserve ammunition in the back of his Pinzgauer. A replacement for the wounded McGhee also descended the ramp of the helicopter. By early evening the BRF were ready to go again, although some rest was desperately needed.

The next phase of the operation was to carry out exactly the same procedure in the next target village. Kughani was a settlement slightly larger in size than Pasab and was about 6km further north. It was centred on another important crossing over the canal which was both wide and fast-flowing. It was clear by now that these crossing points were important to the Taliban and that they were all well defended. That evening the whole BRF once again moved to within striking distance of their target. Observation posts were established in order to see what was happening and to identify any Taliban activity. Everyone else tried to get some sleep but there was great apprehension as to what the morning would bring. In the meantime local intelligence had been gleaned on the Taliban operating in the area around the village. The enemy had anticipated British intentions, they were apparently reinforcing Kughanyi and civilians had been told to get out. Ominously, it seemed that the enemy had a key commander in the area who was organising the operation.

At 0600 the reconnaissance troops moved off in the direction of their objective. They operated in exactly the same way as on the previous day: Holgate's platoon mounted and Seddon's on foot. This gave the vital mix of flexibility, mobility and firepower. Everyone knew that the enemy were ready and most likely waiting for the British to arrive. The dismounted troops expected to be in contact at any time, but surprisingly it was the BRF that drew 'first blood'. A six-man team commanded by Sergeant Barrow of the King's Royal Hussars moved into the outskirts of the village and to their amazement stumbled right across the enemy. Three fighters were lying behind a shallow embankment; they were clearly waiting in ambush positions but had not seen Barrow and his men, who had approached from their flank. The three Taliban were less than 20 metres away, and it was only a matter of time before one of them glanced to his side and spotted the BRF men. Barrow and his troops opened fire and killed all three of the would-be ambushers. The sound of gunfire alerted every enemy fighter in the area and suddenly everyone seemed to be in contact. The troops already on the outskirts of the village exchanged fire with the enemy at close quarters while the mounted troops hammered away with their big .50 calibre guns. The Taliban were here in force and the weight of fire being exchanged was terrific. Volleys of RPGs were fired from the village and the shooting was interspersed with the sound of the missiles exploding. The dismounted platoon was woefully exposed in this position. The Taliban were able to engage them from the front and from the flanks. Seddon desperately called for more fire support from the mounted platoon.

From the high ground overlooking the village, the mounted elements of the BRF watched through binoculars. Taliban commanders could be seen directing their men and it appeared that they were well dug in and prepared for a fight. Holgate, conscious of the need for maximum fire to be brought down on the enemy, decided to move his WMIK down the slope to coordinate the fire from one of his platoon positions. As he did so an RPG round detonated close by showering his vehicle with shrapnel. Holgate received multiple fragmentation wounds to his arms and legs and the newly arrived driver was wounded in the neck. At company HQ Farrell once again received a request for urgent casualty evacuation. Details of the injuries were being passed by several different sub-units and there was some confusion as to the identity and location of the casualties.

The situation was further confused when Guardsman Stephen Hodgson was shot in the leg. He had been standing in the rear of his WMIK and a bullet had passed through his left thigh. The wounded man sat in the back of the vehicle pressing his hand against the injury. The vehicle commander, Lance Sergeant Lloyd, shouted to Hodgson to use his morphine. Hodgson shouted back, 'No way, that will hurt!', which was a strange comment, considering the circumstances. Frith now moved to the wounded Hodgson in order to evacuate him. In the meantime Farrell had driven to the seriously injured Holgate and had transferred him to the Pinzgauer. The young officer was conscious and coherent but in need of urgent medical attention. Farrell and Frith quickly met up in their vehicles, complete with the casualties, and moved off towards the helicopter landing site. Once again the Chinook with its MERT had already lifted off and was inbound. Farrell's party consisted of three vehicles, which moved at speed to meet the helicopter.

As they drove parallel to the village they were engaged by the enemy and an intense volume of fire was suddenly directed at them. Frith and Farrell realised that they were driving at right angles to the enemy who had concealed themselves in an irrigation ditch about 600 metres away. They now drove through a wall of fire and several volleys of RPG missiles. There were ten or twelve explosions as the deadly projectiles detonated around the speeding vehicles. In the rear Frith shouted into the radio, 'Keep going, don't stop!' He realised that the only thing keeping them alive right now was their speed. The seriousness of the situation had not been lost on Farrell either and he let Frith know that he had absolutely no intention of hanging around. From his exposed position Seddon could now see that RPG rockets were impacting all around and on the raised ground to his rear from which the mounted troops were trying to support him. The volume of fire was incredible, far outweighing that of the previous contacts. The situation was becoming dire and air cover was badly needed.

The helicopter landing site was about a kilometre from the village and Farrell's evacuation party was relieved to reach it in one piece. The Chinook was now seconds away and the casualties were being unloaded when the familiar sound of mortar impacts again sent everyone diving for cover. The deadly bombs fell accurately into the area of the helicopter landing site for the second time in two days. The Taliban knew the British tactics and that a helicopter would be arriving. The BRF were learning too and they looked to identify the mortar fire controller. Sure enough the enemy fighters were using a mobile platform to direct the fire and after positively identifying the controller, he was neutralised by BRF fire. The Chinook had now arrived but had seen the mortar impacts in the area of the proposed helicopter landing site. It would be folly to land there so Farrell once again moved to a new location and signalled to the helicopter, which promptly landed close by. Holgate, now bandaged, was carried between Farrell and Tindle who linked arms to support the wounded officer. Hall hoisted Hodgson over his shoulder and carried him toward the waiting helicopter. As he climbed a steep bank on his approach, Hall lost his footing and slipped dropping Hodgson face first into the sand. Time was of the essence so Hall once again hoisted his wounded comrade onto his back and completed the journey to the aircraft. Hodgson joined Holgate on the floor of the Chinook and Hall triumphantly rejoined Farrell. The Chinook lifted off with the medics already getting to work on the wounded men. The wounded replacement driver courageously remained in the fray and his injuries were, fortunately, not too serious.

Major Sergeant had by now assessed the enemy's strength. The Taliban were much stronger here than those encountered over the previous two days and were better organised and led. Their positions were well prepared and they were heavily armed. There was nothing further that the BRF could achieve by closing with the enemy and the extraction would be very difficult. The Apaches had arrived once more and were now able to add to the enemy's problems, but intense fire was still raining down on the dismounted troops. It would be impossible to move back over the open ground without taking further casualties unless the Taliban could be silenced. 'Fast air' in the form of two US Air Force F16s arrived right on cue. They were directed onto several areas where the enemy had been identified and the troops were able to hear the reassuring roar of jet engines once again. Three 1,000lb bombs were dropped to devastating effect. The ground shook and grey smoke rose from the now flattened enemy positions. This was Seddon's opportunity and he ordered his exposed platoon to withdraw while the Taliban were stunned and engulfed in thick dust. They began their rearwards move firing all the time in support of each other. The enemy returned fire and as they recovered from the shock of the aerial attack, this intensified. It was exhausting work for Seddon's platoon but they managed to get out of the killing area without taking any further casualties. The British troops withdrew in sequence, all the time hammering away at the enemy. The Taliban gave as good as they got and did their best to inflict as much damage as possible with their RPGs. The Apaches were able to use their rockets to cause further confusion. The BRF moved back in good order. The contact had once again lasted about five hours and everyone was thoroughly exhausted in the heat of the day. Brigade now authorised the BRF to break off contact and to move once again into the desert to another leaguer. Thick smoke hung over the village and the chain guns from the 'Uglies' could be heard from some distance.

It had been a hell of a fight. There could be no doubt that scores of Taliban had been killed during the day, many of them buried under the rubble of their defensive positions. The BRF had lost another two men to wounds but this aside the mission had been a success. The enemy had been drawn out, their routes and areas of concentration had been identified and the brigade HQ staff could now identify more clearly their targets for the forth-coming Operation Silicon. The BRF were tired and low on ammo. Brigade gave the order for them to return to Camp Bastion. As they drove through the desert, the troops were conscious of what they had experienced in the last three days. They had taken four casualties of their own and had a full six months ahead of them. It was a sobering thought and although they didn't yet know it they hadn't seen the last of Kughanyi either.

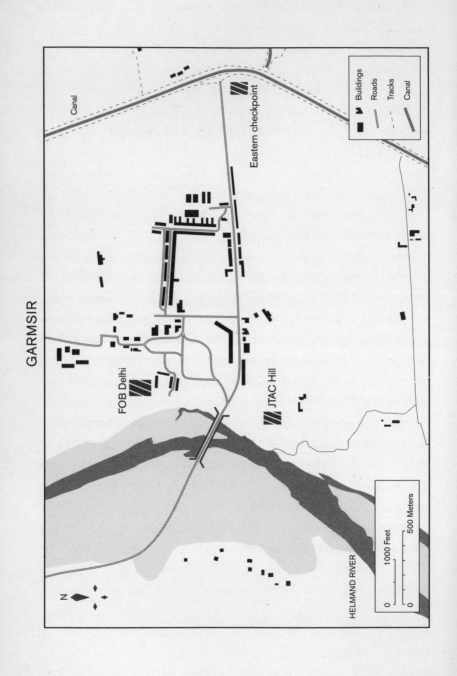

GARMSIR

Canal

Eastern checkpoint

FOB Delhi

JTAC Hill

Buildings
Roads
Tracks
Canal

N

HELMAND RIVER

1000 Feet

500 Meters

0

0

8

GARMSIR AND
3 COMPANY

The main body of 3 Company started their move to the southern town of Garmsir on 19 April. Major Will Mace and the reconnaissance party had flown in some days earlier and were getting a feel for the area from the outgoing Marines. He had taken most of his officers and company HQ with him, so CSM Robinson was left to organise the move of the company from Bastion. There was a buzz of excitement around the troops; at last after all of the waiting they were deploying. 2 Platoon under Lieutenant Jeremy Quarrie had already been sent on a short attachment to Now Zad and they were to join the rest of the company later. CSM Andrew Robinson shepherded the troops as they struggled under the weight of their equipment and the additional stores they were carrying with them. The heavily laden troops crowded into the interior of the Chinook and it lifted off for the uncomfortable journey. It was about 100km to Garmsir and the flight time was about half an hour. There was very little of note for the passengers in the vibrating helicopter to see out of the small portholes, just miles of desolate desert. No one quite knew what to expect when they arrived and there was a little apprehension in the air.

The aircraft soon started its descent and some of the troops on board were surprised to find that they were not landing in a secure base area but were putting down in the desert. They seemed to be miles from anywhere. Huge clouds of dust obscured everything so that goggles had to be worn to unload the large quantities of equipment that they had brought with them. A couple of vehicles could be seen through the dust but the landscape was otherwise bare. It took some minutes to unload the aircraft, but eventually the powerful helicopter lifted into the air and sped off in a northerly direction. It was too dangerous to land the fully laden aircraft near the town as it was in range of Taliban weapons. Landing sites were regularly selected and secured in the expanse of desert a few miles outside the town and away from the enemy. As the rhythmic beat of the rotor blades steadily disappeared, Robinson introduced himself to the Marines who had been sent to receive them. It was clear to Robinson that there was nowhere near enough transport to ferry all the troops and their stores in one go. A large civilian dumper truck seemed to be the only significant transport available, so the NCOs organised the loading of the stores and then the troops onto the truck. It was ridiculously overloaded but made the two journeys necessary to move the Grenadiers into Garmsir. The Guardsmen were very surprised at the appearance of the Marines when they got there. Many wore thick beards with long hair and all of them seemed to be very thin. Their clothes were dirty, torn and they were without exception browned by the sun; some of them wore shorts. Weeks earlier the Marines had fought to capture the town from the Taliban and they had clearly had a tough time of it. The Guardsmen wondered what was in store for them.

FOB Delhi would be 3 Company's home for the foreseeable future. FOB Delhi was centred on an old agricultural college, the centrepiece of which was three very basic single-storey, flat-roofed buildings. The Engineers had constructed a perimeter from sand-filled bastion walls. These were grey wire baskets and seemed to be the main feature of any deployed accommodation. They were simple, practical and provided good protection from enemy fire.

The walls were two baskets high and surrounded the buildings within, making the whole compound about 100 metres by 200 metres square. There were some smaller buildings and one of these was utilised as company HQ. Two 20-foot containers had been positioned inside the compound. One of these was for the storage of rations and the other for stores. There was also a refrigerated container but this had not been working for some time.

CSM Robinson was surprised to find that there were no gates in the perimeter, just a gap in the walls. He made a note to ensure that this was properly defended. The interior of the compound was ankle deep in fine dust and there were bullet holes in the cement walls of every building. There was also evidence of RPG strikes on the bastion walls and sand slowly seeped from the holes in the torn material. A single generator provided enough power to light two of the buildings and nothing else. On each corner of the base, on top of the flat-roofed buildings, sangars had been constructed from bastion walls and sandbags. An array of weaponry could be seen in the sand-filled constructions and these positions were obviously key to the defence of the FOB. To the west side of the compound the fast-flowing Helmand River passed within 50 metres or so of the base and this feature acted as a natural barrier. The base was approached in this direction by means of a steel girder bridge over the river which was of obvious tactical importance. To the north the high-walled compounds of the town spread out to the fields and irrigation ditches that surrounded the entire area.

For now the Grenadiers would have to concentrate on the interior of the base. Time was, as usual, of the essence and the Marines were understandably keen to go home. The advance party were quick to receive their troops and to get them organised. The new arrivals were surprised to find that the appearance of their own officers and NCOs had also changed. Guardsman Swann barely recognised Lieutenant Andrew Tiernan, who was now sporting a beard and a significant tan. The officers and NCOs of the advance party had already seen some action and were starting to resemble the Marines they were to replace. Briefings were delivered and the men slowly became familiar with their surroundings. Equipment was, as always, in short supply; there weren't even enough camp cots to go around and some of the men had to sleep on the floor. The CSM and those responsible for administration set about taking over from the Marines, but were surprised to find that there appeared to be no form of accounting in place. The Grenadiers simply took what they were given. There was some isolated accommodation for Marines with the diarrhoea and vomiting bug. Ominously, there were a good few sick men residing there.

In the following days the Marines thinned out and eventually 3 Company was left to its own devices. The container in which the rations were stored was opened up and CSM Robinson discovered that it was rat infested. The loathsome creatures had been feasting on the compo rations for some time and about half of the food had to be thrown away. He also did a 100 per cent check of all the equipment in the base as there was no record of what was being held. Much of the inherited ammunition was in a rusty condition and a request for replacement stocks was quickly sent; the new arrivals would soon discover that ammunition expenditure in Garmsir was significant. Vehicles, too, were in short supply. There were only six WMIKs; two of these were unusable and none had armour. In the following days the troops set about cannibalising and rebuilding the vehicles under the supervision of Lance Corporal 'Dagger' Dawson.

There was a medical officer in Garmsir and he was equipped with the basic life-saving materials that were required in such an isolated outpost. He operated from a small room which served as the company aid post. Everything that happened in the area was controlled from the company operations room, which was little more than a small room with a map table in the middle.

Empty ammunition crates were used as desks and maps were pinned to the walls and spread over various surfaces. From here Mace commanded the company and Captain Rupert King-Evans directed the daily operations in the isolated outpost. Everyone quickly settled into a routine as they became familiar with their new home. For most, the worst aspect of Garmsir was the food. There were no fresh provisions and the compo rations quickly became monotonous and for some unpalatable. Corned beef hash seemed to be the staple diet and many simply couldn't face it day after day. In the increasingly hot conditions it was also important to drink plenty of water. Ideally each man needed to consume several litres of liquid a day. The troops were allocated only two bottles of drinking water each per day. There was no way to cool the bottles so the liquid was usually warm and disgusting to drink. The remainder of the daily required intake came in the form of purified water drawn from a well. This was entirely safe to drink but tasted like warm swimming pool water. Most of the troops used the dreaded army orange juice powder or 'screech' to make the liquid drinkable. Screech was not always obtainable and the men even resorted to using hot chocolate powder to improve the taste.

When the Taliban were forced out of Garmsir they did not travel very far. At this time the town marked the furthest expansion of government control to the south. For this reason the enemy wanted to keep pressure on the soldiers holding the small urban area. Being so far to the south, resupply and reinforcement were always difficult and resources could sometimes become stretched. The Taliban were aware that the coalition soldiers inside FOB Delhi were quite isolated and so they dug themselves in on the south and east of the town in order to continue their attacks. For the most part, the area 1km to the south of the base was clear of enemy fighters, although they constantly tried to infiltrate to the north. A ten-metre-wide canal running to the south-east provided a useful obstacle upon whose banks the British could mount standing patrols and checkpoints. It was important to hold the Taliban line back in order to protect both the town and FOB Delhi. The resultant stalemate was reminiscent of the Western Front with neither side able to completely vanquish the other. On the south-eastern side of the canal the Taliban had dug a complicated network of trenches and rat runs. Ruined buildings and compounds had been prepared for defence and the enemy was able to move around largely concealed from British observation. They used these trenches and tunnels to infiltrate forward in order to attack the British positions at every opportunity.

The ground to the south of FOB Delhi was similarly fortified by the enemy. The British, who were constrained by numbers and resources, were forced to observe these areas and, where the enemy was seen, to direct the most effective weapon systems onto them. 81mm mortars could be fired from within FOB Delhi and if necessary 105mm light guns could be called upon from FOB Dwyer some 10km away to the west. The various enemy areas were divided by the Grenadiers into zones each named after girls or drinks such as 'Kylie', 'Britney', 'Christina', 'Vodka', 'Scotch' and 'Whisky'. Among the ruined and cratered landscape in these areas were reference points which were quickly identifiable. 'Frog Eyes', 'Five Cigars' and 'Two Trees' were such areas and the Taliban were often engaged in them. Fast jets had regularly been called for support to the Marines and the whole area especially to the south of the canal was pitted with bomb craters and shell holes, adding further to its similarity to a First World War battlefield.

To deter the enemy from moving up from the south and east, two permanent checkpoints had been established. In the west on the banks of the Helmand River, 'JTAC Hill' overlooked the waterway and provided good all-round views of the approaches to the town. It dominated the ground to the south and its raised position allowed good observation for a considerable distance. The checkpoint on this small knoll was protected by the inevitable bastion walls and many sandbags. The little redoubt bore the scars of frequent attacks and its tactical importance to the British had not been overlooked by the enemy, who used every opportunity to harass it with both direct and indirect fire. About 1km to the east a second checkpoint, known as 'Balaclava', had been established. It lacked the dominating position of its neighbour to the west but was situated right on the western bank of the canal which provided an effective obstacle to the enemy. Balaclava was an empty shell of a small cement-walled building. Some observation of the ground to the east and south was afforded from the roof and this had been reinforced to take the weight of the sand-filled defences now in place there. A single aluminium ladder inside the building provided access to the rooftop observation post. A very small protective bastion wall surrounded the little house and it had taken a severe battering from enemy fire. The outside of the building too was pockmarked with many bullet holes and a little barbed wire had been added as a final deterrent to the enemy across the canal. The eastern checkpoint had a reputation as a dangerous place to be as this was regularly attacked and the enemy were able to get very close. The rugged ground and thick vegetation on the far side of the canal sometimes allowed the Taliban to get within grenade-throwing distance and the fighting here had been both close and intense.

A routine was established in FOB Delhi which saw the various platoons rotating through a series of mainly defensive tasks. A guard force was required to secure the base and a section at a time was allocated to each of the checkpoints, usually for around 24 hours. The specialist snipers and Javelin operators were allocated to a different rotation to ensure their availability when needed. A couple of 81mm mortar sections operated from pits at the northern end of the compound. These men were frequently called upon to provide supporting fire to those exposed outside the FOB. They had to react at a moment's notice and could often be found firing the mortars half-naked. There were some local patrol tasks to be carried out and reassurance visits to the nearby police detachment were required daily. Contact with the local population was important as this often yielded intelligence on the enemy. It was likely that the local farmers were playing both sides off against each other and they clearly couldn't be trusted, but it was essential to maintain a good relationship with them. Regular reconnaissance and fighting patrols were mounted in order to keep the enemy off balance and these were usually conducted during the hours of darkness.

As a result of their ability to lose their weapons and revert to being 'civilians', the Taliban were able to infiltrate the town to gather intelligence and to influence the genuine civilians there. They were consequently well aware that the British in Garmsir had changed over and they were doubtless keen to test the new soldiers. After only a couple of days in Garmsir the section on duty at the eastern checkpoint were feeling a little isolated as darkness fell. The small building was a frightening place to be when everyone knew the enemy was capable of getting so close under the cover of darkness. After several hours the soldiers had started to relax when some local intelligence was received. The interpreter looked very worried as he quickly relayed the information in English. According to his translation the Taliban were closing in and were about to mount an attack on the little building. The outpost was apparently surrounded and an assault was imminent. The troops stood to and prepared for a ferocious battle. From the rooftop they peered into the darkness and strained to hear any sign of movement. Nothing came. After a long and nervous night the Grenadiers realised that they had been the victims of a Taliban mind game. A couple of fighters had caused great alarm to the new arrivals and were probably sitting in the trenches nearby laughing about their little prank. They would not have been able to see the effect of their ploy, but they certainly wouldn't have missed the Javelin missile that Guardsman Hennell fired in the early hours. It may have been a nervous night but the fighting capability of the new arrivals would not have been lost on the Taliban.

Attacks on both checkpoints soon became commonplace and very often they were engaged three or four times a day. The enemy were usually seen as they attempted to get close enough for a good shot and the assault would be brought to a close by overwhelming fire from the .50 cals and GPMGs. The enemy strength varied from single fighters attempting to snipe those on view in the two checkpoints, to larger groups mounting determined and sometimes close quarter attacks. The mortars fired regularly from the pits within FOB Delhi and the impressive firepower from the 105mm guns was frequently used. Daily sorties over the southern town were called to drop tons of ordnance on the attacking Taliban positions and the British troops were grateful for the air support they received. Scores of enemy fighters must have been killed in this way, but their replacements kept on coming. In these early days 3 Company were focused on familiarising themselves with the ground and the enemy method of operations. It would be folly to mount any large-scale offensive operations until the company was properly consolidated in FOB Delhi. It took a while for the relief in place to be completed and for 2 Platoon to rejoin them from Now Zad. Patrols were quite small-scale and were for the most part focused on the gathering of intelligence and on ensuring the security of the town. Several attacks a day kept company HQ busy calling for artillery or air support. The intense heat, primitive conditions and frequent enemy attacks ensured that everyone became tired very quickly. The eastern checkpoint was identified as the most dangerous location and at the beginning of May some of the men noticed that the Taliban seemed to be using different ground to mount their attacks. Most activity had been to the south in area 'Vodka' but they now seemed to be attacking from the east as well. They seemed to be pushing closer and were using more RPGs. The enemy were often seen moving into position and were effectively engaged. Once they started shooting, the contacts usually lasted from half an hour up to several hours. They would often break off contact and then after a short pause would start shooting again. A favoured tactic was for the Taliban to sneak into their firing positions during the hours of darkness and then to open fire just as the sun came up behind them making the attackers difficult to spot.

On the morning of 3 May a section from 1 Platoon was preparing for a handover of the eastern checkpoint to their relief which was due later that morning. It had been a long night of observation and everyone was tired and ready to return to the relative safety of FOB Delhi. The men had packed up their kit and were now setting about the normal routine, conscious that this was a high threat time. Swann was on duty behind the GPMG and was peering into the morning light for any signs of the enemy. Guardsmen Parker and Simon Davison were working on the roof preparing equipment for the handover. At around 0645 hours, Davison relieved Swann so that he could pack his equipment downstairs. The remainder of the section were inside the building. Swann was aware that those on the roof had requested a UAV flight as they had apparently seen something suspicious. Suddenly the familiar sound of the rooftop GPMG blasting away was heard and everyone reached for their weapons. The checkpoint had come under attack from a number of enemy positions and the sandbagged emplacement was raked by automatic fire.

Davison blazed away with the machine gun as his comrades joined him on the roof and also returned fire. As they fired, a shout of 'man down' was heard; the GPMG had stopped firing and a quick glance showed that Davison had been hit. He had been shot in the head and his injuries were severe. Colour Sergeant O'Halloran, the CQMS who had been visiting the building, helped by others, applied a field dressing and gave immediate first aid. The attack was continued and those not involved in treating Davison returned the high rate of fire being directed against them from across the canal. Back at FOB Delhi the contact had been heard and news of the casualty at Balaclava had reached the quick reaction force (QRF) who now raced in their vehicles to support the beleaguered section. Robinson together with the medical officer, Drum Major Matthew Betts, and his platoon commander, Lieutenant Stuart Jubb, joined the race to help. The journey to the checkpoint was brief and as they arrived the QRF could see tracer streaking across the roof of the small building.

O'Halloran was now firing the 51mm mortar at the attacking Taliban and the QRF quickly took up fire positions in support. Robinson ran into the house and was told that the desperately wounded Davison was on the roof. Above them Swann and others fought to save their friend's life. Moving the wounded man down the aluminium ladder was a difficult task but Robinson hoisted the young soldier onto his shoulder and climbed down the ladder until he could pass Davison to Betts and other members of the platoon who were waiting beneath. As the wounded man was placed on a stretcher the medical officer assessed his condition and further medical aid was given. Shooting continued all around as the casualty evacuation continued. Davison's condition was critical and no time was wasted in placing him onto the waiting QRF vehicle which raced off to the helicopter landing site from where he could be flown to the hospital at Bastion.

At Balaclava the battle continued even as the Chinook arrived to extract Davison. Air support had been called for but this would be tricky. Some of the enemy were quite close and the air strike would have to be brought in 'danger close'. This meant that bombs would be dropped at the absolute minimum safety distance for friendly forces and the margin for error was very slim. Swann, who had moved to occupy a ragged sangar about 50 metres away, now saw the ground from a different angle and as he searched for a target an enemy fighter obligingly showed himself at a range of about 300 metres. Swann fired and the Talib went down. 'Fast air' arrived overhead and after some liaison with the ground controllers a series of bombs was dropped. The immediate landscape was transformed into a swirling dust cloud and the shockwaves shook the whole building. The fight was brought to a close as the Taliban were either killed or once again forced away.

Sporadic fire continued throughout the day and the British troops directed mortars onto suspected enemy positions. This had been a typical contact, but Davison was the first casualty the company had taken. It was obvious that his injuries were severe but everyone prayed that he would make it. Davison had continued to man his position on the roof and had provided covering fire while his comrades got into position beside him. Later that day, Mace had the unenviable duty of informing the company that Guardsman Simon Davison had, sadly, not made it. The 22-year-old had joined the 1st Battalion a little over a year previously. He was the first Grenadier to be killed in action since 1993 and his loss was a terrible blow. There was a great air of sadness in Garmsir, followed by an intense anger which was directed towards the fanatical enemy fighters hiding in the ruins to the south and east. A grim determination to pay the Taliban back tenfold soon gripped the company. For now the grieving process would have to wait. A tremendous fighting spirit developed and the Taliban were soon attacking again, providing ample opportunity for payback.

While the company had been consolidating and learning the ground, Mace, the company commander, had been planning a series of offensive operations. The most obvious way to push back the forward line of enemy troops was by mounting aggressive fighting patrols. In this way the Taliban would be harassed out of their hiding places to the south and their freedom of movement could be restricted, particularly at night. Mace was doubtless aware of the need to maintain an offensive spirit. When soldiers were confined and subjected to repeated attacks, morale could easily dip unless carefully managed.

A couple of days after Davison's death 3 Company moved out under the cover of darkness to try and surprise the enemy. During the day detailed orders had been given and rehearsals took place. Mace's plan involved crossing the canal by placing a lightweight infantry bridge over it. This small aluminium construction could be carried by a section of troops and placed quietly over a gap. Additional training on the narrow footbridge was required so that it could be put into position silently and efficiently in the dark. The rehearsals took some time and the placing of the bridge was repeatedly practised until everyone was familiar with it. Once across the canal the company would move through largely unknown ground and mount an ambush on firing positions that had been regularly used by the enemy.

In the early hours of the morning the company patrolled out from FOB Delhi and headed in the direction of the eastern checkpoint. The men manning Balaclava were able to provide additional support and they scanned the area across the canal for any sign of movement while the flimsy metal bridge was placed. It was a difficult and time-consuming business to get the structure safely in place without making any noise, in the dark. It took almost two hours from leaving Delhi until the bridge was in position and Mace ordered his lead troops across. Only one man at a time was able to cross the bridge because of its flimsy construction and it took a long time to get everyone quietly over. They were now in unknown territory. No one had been this side of the canal and the enemy could be anywhere. The troops were heavily laden with body armour, helmets and large quantities of ammunition. The whole area was a mass of bomb craters and ruined walls which were crossed by ditches and trenches; it was a nightmare to negotiate. The thick undergrowth further complicated any movement. Each soldier used the infra-red monocle mounted on his helmet to find his way in the darkness. When looking at the ground through the IR device, there was no depth perception and it was very difficult to safely place one's feet. The uneven ground and the pitch black of the night made progress even slower as the Guardsmen sweated and stumbled repeatedly under the weight of their equipment.

Eventually the ambush position was reached and the officers and NCOs placed their troops into position. Mace sat with his tactical HQ where he could control the likely battle as it developed. Everyone hoped that the enemy would attempt an early morning attack and that they would be surprised by the now-concealed Grenadiers. The long march into position meant that they didn't have long to wait until first light and as the sun started to come up everyone stared into the mass of ruins and undergrowth. The ground now looked quite different; the area had been devastated by artillery and air strikes and was a real maze. The Taliban had turned this to their advantage and trenches could be seen all over the place. There was no sign of the enemy, the sun was gradually rising and the temperature was already becoming uncomfortable. Suddenly there was a loud 'crump' and then another. An RPG rocket streaked through the sky and exploded nearby. More explosions; these were mortar bombs fired from some distance. It was obvious that the ambush had been compromised and the company would need to extract themselves.

This situation had been planned for and supporting fire from both mortars and artillery was called for. The crews were already standing by and effective counter fire was brought to bear on the Taliban. 3 Company now started a hasty withdrawal under the cover of smoke and high explosives which was keeping the enemy heads down. It was hard work to try and move at any speed over the broken ground, even during daylight, but the troops reached the infantry bridge and steadily crossed onto the friendly side. Enemy mortar bombs continued to fall but were fortunately not particularly accurate. Eventually the last man made it across and the bridge was dismantled. A thick layer of white smoke masked the whole area to the south and east and the crump of high explosive shells could be heard steadily impacting the enemy positions. No-one was injured and the Guardsmen were relieved to reach FOB Delhi by mid-morning.

On the face of things the operation had not been a success, particularly as the enemy had managed to spot the troops. No one knew if they had heard the bridge crossing or whether the Taliban sentries had been alert enough to detect movement. Perhaps someone had just spotted them at first light or maybe they had been tipped off from within Garmsir itself. Whatever the reason, Mace was relieved not to have taken any further casualties. But there were a number of positive points to the operation. Even though the ambush had not been sprung, the Grenadiers had successfully infiltrated an area that the enemy considered their ground. In future they would be much less sure of the British intentions and would have to be more cautious when moving in the dark. 3 Company now had a much better understanding of the ground to the south and of the way in which the Taliban operated there. The rapid response from the British guns had shown that any troops on the enemy side of the canal could be quickly supported. The tired infantrymen had also learned that it was almost impossible to move across the pitted landscape carrying so much equipment. In future much less would be taken, especially the heavy ammunition loads.

In the following weeks 3 Company mounted regular fighting patrols to the south and across the canal. These were often quite successful and the Taliban front line was pushed further back, although they still regularly pushed into the area. Balaclava and JTAC Hill remained the focus for determined daily attacks and the young soldiers soon became battle hardened. They now resembled the troops that they had replaced; many wore beards and most had lost a great deal of weight. The daily firefights with the enemy became almost routine, as did the regular air strikes by NATO aircraft. The Taliban were taking huge casualties as a result of the bombing but they showed no sign of discontinuing their attacks. 3 Company had shown tremendous fighting spirit for a company that had been thrown together at the last minute. This spirit would be maintained for the rest of the tour but they still had some difficult days ahead.

9

CLEARING THE VALLEY

Like Sangin, Gereshk was a key centre of population in the Helmand River valley and it was vital to secure this strategically important town from Taliban influence. Enemy fighters had doggedly hung on to the edges of the urban area and it was now time for 12 Brigade to move them out. The Task Force Helmand plan was to gradually squeeze the enemy out of the populated areas and to replace them with Afghan police and army personnel in order to demonstrate the spread of government influence. Operation Silicon would formally start on 17 April although the BRF and other elements were already working on the shaping phase. It was to be a full task force offensive operation with Battlegroup North providing the bulk of the firepower that would force the Taliban out. A full kandak of ANA troops was to be incorporated and they were to be given their own area of responsibility. Major Martin David and the Queen's Company OMLT now had the task of mentoring 1st Kandak through a large and complex operation. The ANA had not been used in such close cooperation in these sorts of numbers before and the scale of the challenge facing the Grenadiers had not been lost on them. It would also be the first time that the Queen's Company had been on large-scale operations with their Afghan charges so there was a great deal of preparation to be done.

Lieutenant Colonel Carew Hatherley had spent a significant amount of time with General Muhayadin Gori. Each deployment of the ANA kandaks and their supporting elements had been discussed in detail. The Afghan commander was usually keen to commit his men against the Taliban and Hatherley was always available to support him in the planning process. There was a feeling that the ANA could do more and that their British counterparts outside the OMLT did not fully appreciate Afghan capabilities. Hatherley was conscious that confidence had to be built slowly and that this was best done by using Afghan troops alongside coalition soldiers wherever possible. The headquarters element of the OMLT had spent many hours coaching the Afghan staff officers through the battle planning process because only a few of them had received detailed training before.

When the brigade plan was translated into orders, the OMLT companies became responsible for mentoring the kandak officers through their own planning and deployment. Hatherley was very keen to get Muhayadin involved on the ground and for the ANA to be used in a more effective way. The opportunity now presented itself in the form of Operation Silicon. Hatherley's tactical HQ accompanied Muhayadin to Gereshk in order to carry out reconnaissance and detailed planning for the forthcoming operation in the Lower Gereshk valley which would clear the enemy out of the area. Great caution was exercised when the Afghan commander left Shorabak. The Taliban would dearly have liked to kill him; such a high profile Afghan figure would provide the enemy with a significant propaganda coup. The OMLT and the Afghan soldiers were equally determined that he would be well-protected.

David joined his commanding officer on the reconnaissance to Gereshk where he visited the ANA HQ established in the south of the town astride the main highway. Hatherley was keen to establish a joint district command centre here, which would see the OMLT, ANA and the police all working together. Muhayadin was also eager to visit his men in their deployed outposts. The first phase of the forthcoming operation would see the town secured by British and Afghan troops so the OMLT and Afghan officers moved to a large fort in the centre of town from where they could view the area. The old fort had been used as a prison and was quite run down. From the roof David and Hatherley were able to view the Green Zone and the area to the north where their troops would be operating. The Helmand River was a major obstacle but there were a series of canals and watercourses which also broke up the desolate desert landscape. The crossing points over these obstacles would be of obvious tactical importance to both sides. There were scores of high-walled compounds each of which would have to be cleared of enemy. The area of the Green Zone was thick with maize and poppy fields which would provide the Taliban with excellent escape routes. The role of the ANA and OMLT on the operation had not yet been confirmed, but David had a good idea of what they would be asked to do.

As the Queen's Company prepared for their first significant operation, Hatherley decided to drive up to Sangin because there was a major task for the OMLT there too. While the majority of 12 Brigade were still shaking out and learning the ground in Helmand, a major operation had been conducted to clear the Taliban from Sangin for good. The enemy had been in the town for some months and a sizeable force was required to dislodge them. Large numbers of troops from the US 82nd Airborne Division (Task Force 1 Fury) were used along with elements of 3 Commando Brigade who were recalled from Kandahar for Operation Silver. The Americans landed by helicopter to the south and then 42 Commando Royal Marines attacked the town from the north. The Taliban were dislodged after some fierce engagements and by 5 April the Royal Marines were in control of most of the built up areas. British and US troops consolidated in Sangin and the next phase of the operation was put into effect. The US troops took over from the Marines and a British infantry company settled down in the Sangin district centre which had been virtually under siege for some time. The long term plan was for the town to be controlled by the Afghans themselves and the sooner they could be located there the better. While the centre of Sangin was now relatively secure, the outskirts were not. In order to tie down the whole town it had been decided to establish a number of small patrol bases (PBs). These locations around the town were to be occupied by ANA troops under the supervision of the OMLT. The ANA presence would send a powerful signal that the reach of the Afghan government was extending, which in turn would provide some confidence to the population. These PBs would house about a platoon of Afghan soldiers and their mentors. Hatherley now wanted to ensure that these outposts were effectively sited and he sent David by helicopter to liaise with the Americans about the positioning of them.

There were a great number of Taliban still in the villages between Gereshk and Sangin and the journey to FOB Robinson from where the PB construction operation was to be mounted would be highly dangerous. Hatherley had decided that it was necessary to see the ground and to mentor the Afghans through the siting and occupation of the PBs. Hatherley and his Tactical HQ would join a planned logistic convoy to FOB Rob. In the meantime Muhayadin would return to Shorabak and would send his planning staff to FOB Rob by helicopter with David. A couple of days later the OMLT reconnaissance party and ANA troops moved to a desert rendezvous where they waited for the large logistic convoy. When it arrived, the slow-moving procession consisted of British, US and Afghan vehicles. The OMLT CSS troops under Major Andy Parker had arranged a major resupply for the ANA troops operating from FOB Robinson. Several new vehicles were being deployed and the Grenadiers' transport officer, Captain Dave Groom, decided to accompany them. There were a number of heavy trucks brimming with fuel, ammunition and containers full of stores. This was a tempting target for any Taliban in the area if they could get close enough. It was too dangerous to use the infamous Route 611 for the journey and so they headed out into the desert supported by a Canadian Leopard tank and some six wheeled LAV armoured personnel carriers which had arrived from Kandahar Province. The tank led the way as it had considerably more protection from the mines that the convoy was likely to encounter. The Taliban often moved about on small motorbikes which afforded them speed and mobility. They were able to track convoys and to get ahead, placing mines at likely choke points where convoys would be forced to cross obstacles. Since their arrival, 12 Brigade's units had been hit by a number of mines and the convoy drivers were particularly careful to keep their wheels inside the wide tracks made by the tank.

The journey was painfully slow and the drivers of the trucks sweated in their cabs. There was little to see but desert and the route was deliberately remote and away from civilisation. It took more than five hours to reach FOB Robinson and it was early evening when they arrived. FOB Rob was built on a dusty plateau which overlooked both Route 611 and the Helmand River. A wide ring of bastion walls provided the perimeter and it was the best part of a kilometre across the dusty interior. There were a couple of the now familiar flat-roofed buildings and containers but little else. The base was home to a collection of troops, most of whom supported the British company now situated in the Sangin district centre some 10km to the north. A troop of 105mm light guns were positioned here and they regularly fired in support of coalition troops. While FOB Rob provided reasonable views over the surrounding area, it was vulnerable to indirect fire from the Green Zone and the Taliban often fired rockets from the covered area. The OMLT elements of the convoy were quick to locate Captain Ed Janvrin, who had been running his own mentoring team in FOB Rob for some weeks. Janvrin had already seen a fair bit of action and was well versed in the ways of the ANA. Like many of the OMLT soldiers he was more frustrated by the attitudes of the allied troops towards the ANA than the abilities of the Afghan soldiers.

Under Janvrin's direction the various stores were unloaded and he was quick to organise the new arrivals. Groom made sure that the new vehicles found their way to the ANA and that they were secured. Janvrin met with his commanding officer for the first time in a while and was able to update him on recent events in the area. Once the immediate unloading was completed the troops settled down to eat their rations and organise themselves for the night. A short time later there was a sudden whooshing noise and Groom was the first to respond with a shout of 'Incoming!' Seconds later several 107mm rockets exploded in and around the perimeter of the FOB. The only cover available was under the vehicles and everyone crawled into whatever protected space they could find. The attack was short-lived and no-one was injured. As the noise of the explosions faded away, the smell of cordite hung in the air and life slowly returned to normal. This was a fairly routine event at FOB Rob. The Grenadiers dusted themselves off and continued eating their rations. News soon reached Hatherley that their Canadian escorts had not been as fortunate. Having delivered the convoy safely to the FOB, the Canadians turned away for their own base in Kandahar Province. Some time later one of their LAVs struck a mine and there were a number of serious casualties. It was a massive blow to hear such tragic news, but there was nothing to be done.

David had arrived in FOB Robinson by helicopter; he had a number of Afghan staff officers in tow who were to help with positioning the PBs. David discussed the requirements with his commanding officer and the next few days were spent driving around the dangerous Sangin district looking for suitable locations. The Afghans were integral to this process. The siting party started in the north where it was necessary to negotiate with local farmers over the destruction of poppy fields and the occupation of buildings. US troops just to the north were still fighting and a series of air strikes were called for, and it was obvious that the whole area would continue to be contested by the enemy for some time. For now the town was full of patrolling US soldiers who mounted checkpoints on all the possible entry points to the urban area. Once the northern location was sited, the group moved south to repeat the procedure.

At one of the more southern locations David identified a blue house and compound as the most suitable location for a PB. It offered good all-round observation and dominated the area. The ANA officers were less keen on the occupation of this building. Their view was that it was too good a house and would be too expensive. David suspected that there might be other motives and perhaps the owner had some influence. Whatever the reason for the Afghans' reluctance, the coalition troops settled for an inferior but nearby location. Content that the operation was going to plan, Hatherley flew south once again and turned his attention to Operation Silicon. After about ten days in the Sangin area David completed the siting operation and, having confirmed that all the UK, US and Afghan personnel were content, he too left Sangin and flew direct to Lashkar Gah to receive his orders for Operation Silicon. The construction of the PBs started almost immediately and the engineers worked at a terrific rate to complete the job. The Grenadier OMLT would be spending a great deal of time in these fortified locations in the months ahead.

His job at FOB Robinson completed, Groom was dispatched back to Shorabak with the spare Afghan transport. A small convoy of vehicles was to head south and Groom managed to secure a seat in the rear of one of the US Humvees. All of the other occupants were US personnel from the group co-located with the Grenadiers at FOB Tombstone. There were six Humvees and a larger number of mainly empty medium-sized ANA trucks which had been used to deliver stores on the outward convoy. Groom was in the lead vehicle of the convoy which left FOB Robinson for the long dangerous journey south. About 40 minutes into the journey the convoy moved close to Route 611. At this point it was necessary to use the road for a short distance because they were channelled onto it by other obstacles. The long line of vehicles moved cautiously onto the dirt road. This was a very anxious time and everyone was conscious of the recent fate of their Canadian allies.

As they rolled slowly on, Groom noticed that some of the civilians seemed to be leaving their homes. He immediately mentioned this to the US officer commanding the Humvee. To Groom this was a clear indicator that there might be enemy activity planned. The American major scanned the area but no sign of enemy activity was seen and he ordered the driver to move forwards. The convoy now came to another choke point and the driver stopped the vehicle; both driver and commander scanned the way ahead to find the safest route. The US Army officer directed the driver to move forwards and as the armoured vehicle rolled on there was a terrific explosion. The occupants felt the front end of the vehicle rise up with the force of the explosion before crashing back down to earth. The interior of the Humvee immediately filled with dust and smoke as the windscreen blacked out. Everyone's ears were ringing from the powerful explosion. As the vehicle settled back down, the occupants felt themselves for signs of injury and then looked to their fellow passengers. Groom's knees had been forced into the back of the seat in front, they hurt but other than a few bruises he was uninjured.

As the ringing in his ears subsided he heard a cry of pain from the driver seated in front of him, but everyone else seemed to be uninjured. It was obvious to them all that they had been hit by a mine or some other form of improvised explosive device (IED) and there was a great deal of shouting. The front wheel of the vehicle had detonated the device and as they regained their senses attention was focused on the driver who had been seated closest to the blast. The young man held up a hand which was covered in blood and complained that his leg hurt badly. He was clearly wounded, but it was difficult to ascertain the degree of injury from the cramped and smoky interior of the armoured vehicle; no one could reach the man from inside. Groom flung open his door to let out some of the dust and smoke and was immediately subjected to a barrage of objection from the other occupants. It seemed that the US drills in such circumstances were different to those of the British. But it was clear to Groom that the driver needed medical attention. He removed his bayonet from the front of his body armour and adjusted his position so that he could lean out of the door. On the wrong side of 40, Groom wondered what on earth he was doing in this situation, but he used the tip of the bayonet to gently prod for the presence of further mines. The bayonet was pushed into the ground carefully at a low angle so that it would not detonate any of the deadly objects. After about five minutes Groom had cleared a large enough area to stand in. As he looked back along the convoy he noted that none of the other US vehicles had moved. Everyone was locked down in accordance with US procedure which was designed to minimise further casualties. The Grenadier officer now continued the process on his knees, slowly clearing a safe route around the rear of the vehicle. The ground was rock hard and covered in shale which made the job even more difficult. There was no sign of any further mining but he remained conscious that there could be no second chance if the tip of the steel bayonet touched one of the deadly concealed devices at the wrong angle.

This was an agonisingly slow procedure but it could not be rushed and Groom pressed on. Suddenly there was another loud explosion. His heart thumped and he nearly jumped out of his skin; then came relief as he realised that he was still alive and had not been responsible for the detonation. The source of the blast was another Humvee further back in the convoy; the US commander had tried to come around in order to see what could be done to help. The force from a second mine had blown the tyre right off the US vehicle. Groom eventually reached the now open driver's door of the Humvee and he cleared an area large enough for a couple of people to stand in and they extracted the wounded man. The US vehicle commander climbed out over the top of the stricken truck. Groom, now standing next to the injured man, tried to help the soldier from his seat but was greeted with a scream of pain. The driver had what looked to be a compound fracture of his lower leg. Groom took out his own morphine and gave the wounded American a shot of the pain-relieving drug. With difficulty the huge American was extracted from the Humvee and was placed on a folding stretcher.

The experienced ANA soldiers knew how the Taliban operated and had spent months driving over this difficult terrain. In the rear of the convoy they had identified the line of the road that the Taliban had targeted in the mine-laying operation. The Afghans pulled around the stationary US vehicles and drew level, about 20 metres from Groom's position. Groom was able to clear another safe lane to the Afghan vehicles which he thought were probably far enough away from the road to be safe. He marked the route with empty water bottles as he went. The casualty was now moved via the safe lane into the shade of one of the medium trucks. An RAF Chinook was already on the way from Bastion to extract the wounded soldier but now the convoy came under small arms fire from a nearby village. The Americans returned a heavy rate of fire in the direction of the enemy position and the contact was short-lived. Whoever had fired upon the convoy thought better of it.

Before long the Chinook arrived and Groom, the US major from the lead Humvee and two ANA soldiers carried the heavy

American onto the helicopter. A couple of Danish explosive ordnance disposal (EOD) officers walked down the ramp. They had been sent to help get the American convoy out of the mined area and were equipped with some lightweight mine detection equipment. Relieved that the casualty had been extracted and that they now had some expert help at hand, Groom decided that it was time for a cup of tea. He poured out a cup from his flask and shared it with the Americans; he was somewhat disappointed when his allies finished the morale-boosting liquid and returned the empty cup.

As the Danish EOD men went about their business, one of the ANA soldiers pointed out a patch of disturbed ground nearby. This was confirmed as being a further buried mine and it was marked. Groom scrutinised the ground again and spotted a further piece of gravel that looked out of place. It was yet another of the deadly devices. Each of the mines had recent motorcycle tracks nearby, indicating that the Taliban had raced ahead of the convoy placing the mines in the choke point of the road. They had known what they were doing. A series of Soviet-era anti-personnel mines had been laid around the large mine which had badly damaged the lead Humvee. As he surveyed the wrecked Humvee, Dave realised how lucky they had been. The front of the vehicle was completely destroyed and oil from the ruined engine was splashed right across the windscreen.

The US convoy commander was instructed to return to FOB Robinson. They were now short of transport and everyone squeezed into whatever seats were available for the return journey. One of the Danish EOD operators climbed into the passenger seat of an ANA truck. The Americans decided to allow the ANA to lead on the journey back. It was obvious that they were much more aware of the environment than their western coun-terparts and that they would spot something out of place in a more timely fashion. The convoy now headed north once more.

Their direction wasn't missed by the enemy and before long one of the medium trucks also hit a mine. The Danish soldier and the ANA driver were both shaken but unharmed by the very powerful explosion. On this occasion the detonation was followed by machine gun fire. The Americans returned a heavy rate of fire. The ANA acted decisively and rapidly following up into the village, where they arrested six men suspected of being Taliban fighters.

The convoy moved off once again and this time managed to reach FOB Robinson without further incident. It had been one hell of a journey and Groom reflected on the day's events with his American counterparts. The machine gunner in the original Humvee seemed fairly blasé about the whole affair. The British officer heard how this man had survived three separate IED attacks in Iraq. This was his fourth and, it seemed, water off a duck's back. Groom made a mental note not to accompany this man on a convoy again. 2 Company were surprised to see their transport officer back at FOB Rob, but were relieved that he was uninjured. The Americans sang his praises and retold the tale of his courage in prodding his way out of the minefield. He soon achieved cult status with the American detachment.

Back at Camp Shorabak the Queen's Company OMLT group had been preparing for their new and unexpected mission on Operation Silicon. They had frantically tried to acquire sufficient transport for the op, which was always a challenge given the small number of vehicles available to the widely dispersed OMLT. The soldiers were told to be prepared to be deployed for no more than a week on the operation. There was time for the platoon commanders to take their ANA companies on the specially constructed ranges around Camp Bastion and to practise compound clearing drills. Vehicles and personal equipment were packed with ammunition, water and rations accordingly. The available time was spent trying to ensure that the ANA understood what they were required to do and that the Afghan soldiers were suitably equipped for their mission. Experience had shown that there was little or no understanding of logistics, in the western sense of the word. The Afghans would expect to eat fresh rations procured in the field, something the British could not rely on. There was also an expectation that because the British had a superior logistic chain, they would provide the essential items such as water and as a result this was perceived as a British responsibility. CSM Glenn Snazle was at pains to ensure that supplies were available and that the ANA troops had enough essential stores before departure. Even issuing bottled water could be a challenge and it was sometimes necessary to check that it had been evenly distributed. The ANA troops were unconcerned by all the fuss; they had done this all before and they watched with interest as the British rushed around preparing themselves.

David delivered his orders to the OMLT group at Camp Shorabak in late April and he explained the concept of operations to the assembled company. The Queen's Company area of operations would be a strip running to the north-east from Gereshk for about 6km. They were to operate in support of the Royal Anglian battlegroup who would be on their left flank to the north. Both units were to advance north-east and were to clear the Taliban from the Lower Gereshk valley up to a prominent dam at the junction of the Helmand River and the Nahr-e Bughra canal. On reaching their designated limit of exploitation they would assist with the building of a series of PBs similar to those being established in Sangin. From these bases, patrols could then be mounted to discourage the enemy from infiltrating back into the areas they had been cleared from. By doing this it was hoped that the enemy could be prevented from interfering with the reconstruction effort in the important market town of Gereshk to the south. 1st Kandak was to deploy in full and each of its three infantry companies would operate independently under the leadership and supervision of the OMLT platoons. Inside the area allocated to the Queen's Company, the ground was divided into three distinct zones; the thin strip to the north of the canal formed the first of these areas and an ANA company with Lieutenant Paddy Hennessey and his men would work directly with the Royal Anglians here. Across the canal a zone about 2km wide formed a narrow area known as 'the finger'; Lieutenant Will Harries' platoon with their Afghans would clear this. The finger in turn was bounded by a tributary of the Helmand River on its south side. On the other side of the river the third Afghan company would push north towards the village of Zumbelay with 3 platoon under Second Lieutenant Folarin Kuku in order to occupy any enemy forces there. In effect, each company was separated by a strip of water. It was clear to David that this could make command and control quite difficult and he would have to rely heavily on the leadership of his young platoon commanders. The ANA were to operate effectively on the southern flank of the UK battlegroup and some suspected that the British battlegroup wanted to keep the Afghans out of the way.

In military operations it is critical that all elements of a deployed force are in position before anyone makes contact with the enemy. Much coordination and preparatory movement are required to ensure that everyone is ready to go on time. This movement is usually quite a complex operation and is prone to setbacks such as vehicle breakdowns and other unforeseen events. The Royal Anglians were not keen to take the Afghans with them on their move through the desert because it was felt that this might draw attention to them and compromise the operation before the line of departure (an imaginary line denoting the start of the operation) was crossed. Therefore it was decided that the Grenadiers would lead the Afghans into position. The Queen's Company were understandably concerned that the ANA would not be in position in time. Afghan time-keeping was very different to British and as H-hour had been set for an early morning start, it was decided that the Grenadiers would lead 1st Kandak into concentration areas closer to the start line on the afternoon before the op was due to begin. Padre Stephen Dunwoody conducted a brief field service for the company before they mounted their heavily laden vehicles and moved off to meet up with their Afghan allies. To the utter amazement of the Grenadiers the Afghans were ready to go, their vehicles were packed and the ANA soldiers looked relaxed and unconcerned as they puffed away on their cigarettes.

The convoys departed leaving huge dust trails behind them. Within half an hour, two of the British vehicles had broken down and had to be towed into FOB Price for repair. To compound the problem the radios had packed up too. The Grenadier NCOs were furious and a little embarrassed that they were having to use equipment that was less than perfect. The ANA seemed to have no such problems with theirs. Amid much shouting and swearing the mechanical and radio problems were resolved and the troubled vehicles were able to rejoin their convoys a little later on.

Each Afghan company was headed for a slightly different area adjacent to their given start line so that they were ready to move straight to their axis of advance in the morning. The dominant ground in the area was a low feature known to the troops as 'Hamilton Hill'. The BRF were to secure this feature and hand it over. The Queen's Company HQ under David would establish itself there during the early stages of the op. Captain Rob Worthington, the battalion mortar officer commanding the Fire Support Group would coordinate the use of all the heavier support weapons and Sergeant Dan Moore would set up two 81mm mortar barrels so that he could bring fire to bear on the opposition should the need arise. Each of the Grenadier platoons together with their ANA companies used the late afternoon to move towards their designated areas.

No sooner had they reached the southern outskirts of Gereshk than Harries and his party were suddenly ambushed from prepared positions on the other side of the canal. The convoy immediately stopped and returned fire. Hennessey's group were simultaneously attacked on the north side of the canal. Hennessey was concerned that in the confusion he could be engaged by Harries' group across the water. The enemy were no more than 50 to 100 metres away from Harries' troops but were separated by a formidable water obstacle. The ambush had been initiated by an RPG rocket which exploded nearby. As he watched from his vehicle, CSM Hill saw a second RPG pass between his own vehicle and the ANA truck directly in front of him. Things were becoming distinctly unhealthy. The .50 calibre machine guns on the Grenadier WMIKs were quickly brought to bear on the compounds and Lance Sergeant Ball blazed away with his heavy machine gun from the rear of the platoon commander's vehicle. The ANA troops fired wildly across the canal at the Taliban positions. Sergeant Clint Gillies and the other British NCOs tried desperately to control the ANA fire from behind a small embankment where most of the troops had taken cover. The noise was now quite deafening but it was becoming clear that the firefight was being won by the OMLT and ANA. The brown compounds on the other bank were now a haze of dust as the heavy calibre bullets from the .50 cals tore into the walls. Hennessey decided to dismount and to push up on foot with his Afghan troops to the hydroelectric dam ahead. Harries agreed to do likewise on the south side of the watercourse and the two young officers maintained close contact on their radios.

Accompanying Hennessey's group was the ANA Support Company Commander, Qiam, or 'Rocky' as he was known by the Grenadiers. Rocky was infamous for his fearless and aggressive attitude towards the Taliban, which often bordered on the downright reckless. His father and brother had been murdered by them during their reign of terror in the country prior to 2001. Hennessey's group soon came across an Afghan National Police bunker which was positioned to defend the approaches to the hydroelectric dam. On entering, the inhabitants were found lying on the ground and making no attempt to influence the battle in any way. Qiam screamed at them but the policemen were more interested in complaining about their lack of body armour than in fighting the enemy. Meanwhile, Lance Sergeant Rowe and Lance Corporal Price took up firing positions and with their ANA troops attempted to suppress the forward enemy positions. Qiam was as usual desperate to kill Taliban and he rushed forwards alone across open ground to close with them. Hennessey screamed after him but to no avail. He was now 50 metres into an open field in the middle of a gun battle with no cover. Hennessey and Price managed to eventually drag him back with fire support from Gillies and Harries who had now pushed up to join them. The enemy fire soon ceased completely but it took some time to control the ANA who seemed intent on killing everything in view. The Grenadiers eventually managed to get the Afghans to cease firing and it became clear that the enemy had broken contact and slipped away.

The BRF secured the feature known as Hamilton Hill which was duly occupied by David and the Queen's Company HQ. Worthington looked for the best positions for the Fire Support Team with their heavy machine guns and Javelin missile launchers; under his supervision mortar pits were dug and Moore set up his 81mm mortars alongside the Royal Anglians' mortars which were co-located with the company for this phase. He felt sure that they would be needed. Before the night was out Harries and the rest of callsign Amber 61 were attacked again. Yet, in spite of these early engagements the whole of 1st Kandak reached its designated area according to plan. However, there was sporadic firing throughout the night and few people managed to get much sleep.

A violent dust storm made life even more miserable for the British troops. No casualties had been taken and David was fairly pleased with what had been achieved with the ANA so far. H-hour was set for 0315 hours and the OMLT shepherded their respective kandaks into position in good time. Hennessey's Amber 63 were on the left, across the canal. Will Harries and his men prepared to advance up the 'finger' and Kuku prepared Amber 62 to tackle the right-hand sector. Worthington moved down from the hill to a position on the finger where he was able to assist with coordination between the platoons and company HQ.

At H-hour the coalition troops started to advance in a northerly direction. Any resistance was to be dealt with and all Taliban were to be cleared from the area; those wishing to remain would die where they fought. David remained on Hamilton Hill coordinating the efforts of the OMLT troops and discussing options with the battlegroup commander.

In the southern area, Kuku's men were frustrated by an enemy that used delaying tactics. The Taliban repeatedly engaged the advancing troops, withdrawing to new positions when things got too tough. They were slowly drawing Kuku and his men towards Zumbelay. During one particularly vicious contact the enemy tried very hard to surround the advancing coalition troops and disaster was only averted by conducting some determined counter-attacks. Worthington, the battalion mortar platoon commander, watched the fall of shot as Moore's mortars fired. On several occasions the heavy weapons provided very welcome support to the troops further ahead. Snazle organised the company headquarters and prepared to ferry ammunition forward if required. For much of the day those on the hill could do little but watch the dust rising into the air from the explosions in the distance. Snazle was joined by CSM 'Des' Desborough of the Parachute Regiment, who had volunteered to join the Grenadiers in Afghanistan. As the two warrant officers drank tea on the hill, there was a terrific explosion close by, quickly followed by another rather closer detonation. The tactical importance of the hill had not been overlooked by the Taliban who had rightly identified Worthington's mortars as being a significant threat. The explosions were Chinese-made 107mm rockets, which continued to arrive at an alarming rate and crept closer up the hill. Snazle and Desborough were reluctant to spill their tea and so they hunkered down in some old Russian trenches as pieces of lethal shrapnel cut through the air above them. Both men were experienced soldiers and they continued their conversation between the loud crumps from the rockets; they took great care not to waste their precious brew. Desborough had experienced enemy artillery before: 24 years earlier he had sought cover among the rocks of the Falkland Islands with 2 Para. During the height of the bombardment an Afghan soldier came sauntering over the hill, seemingly oblivious to the maelstrom around him. A rocket exploded only about six feet away from him, but incredibly he was unharmed, although a little deaf for a while. Fortunately the rockets soon ceased and no one was injured. Coalition aircraft dropped bombs onto the enemy and the little group on the hill had a grandstand view of the action.

Kuku's callsign Amber 62 continued to meet with resistance on the outskirts of Zumbelay. The BRF had moved around to the north-east and had already identified a number of enemy positions. After some small arms fire the enemy bunkers were eventually destroyed by the use of Javelin. The Taliban, undeterred, replied by firing rockets towards the BRF positions and this pattern continued throughout the day. Although the BRF had identified what they believed to be the enemy HQ in the village, they were hindered by the presence of civilians and remarkably by groups of children playing in the area. Meanwhile, Kuku's men were now in heavy contact. The callsign was exchanging fire with an estimated 20 Taliban and were increasingly being pinned down.

During the battle Lance Corporal Jack Mizon demonstrated the extreme courage that was to become his trademark during the tour. The enemy brought a huge weight of fire to bear on the OMLT and Afghan troops, using RPGs, and medium and heavy machine guns. An hour into the battle, ANA resolve was starting to crumble and they threatened to withdraw unless more heavy weapons could be brought to bear on the enemy. Mizon, who was deployed with the lead platoon, used his own initiative and sprinted back 800 metres to the vehicle muster point. He was in view of the enemy throughout and in full knowledge that he presented a tempting target. Once there he took a GPMG from its vehicle mount and gathered as much ammunition as he could carry. Under heavy machine gun and RPG fire he returned to the lead platoon where he quickly used the machine gun to put down a heavy and accurate rate of fire onto the enemy. The ANA, injected with new confidence by Mizon's bravery and targeted fire, now surged forwards and dislodged the Taliban from their strong points. David requested fire support from the BRF who provided it from the flank; this proved to be decisive in allowing Kuku to extract his men from their precarious position. During the battle, one Afghan soldier was hit and seriously wounded. The Grenadiers helped to treat the injured man and the MERT soon arrived in a Chinook to evacuate the casualty. Prompt action by the Grenadiers saved the ANA soldier's life, a fact duly noted by the Afghans who were also impressed that the British casualty evacuation system worked so efficiently.

On the northern side of the canal Hennessey's Amber 63 advanced on the right flank of the Royal Anglian's B Company. They effectively operated as a mobile reserve for the dismounted British infantrymen who struggled through the tough and dangerous Green Zone. Amber 63 cleared a series of compounds but encountered little opposition. To the south, Harries and Gillies advanced slowly along 'the finger' with their group but there was no sign of enemy activity in this central area. Some time after mid-day, B Company came under sustained Taliban fire; their mobility was very limited as vehicles could not be used in the Green Zone and the young soldiers were heavily laden with ammunition and equipment. The enemy rate of fire was quite intense and it was difficult for the Anglians to close with the enemy without taking heavy casualties. To compound the problem enemy mortar rounds started to land nearby. The B Company commander formulated a plan and called Hennessey forward to brief him.

The OMLT and ANA were still mounted in their vehicles and so Hennessey was quickly able to move up and join B Company. Too late, he realised that he had taken his WMIK a little too close, as enemy rounds started to strike nearby and a couple ricocheted off the vehicle's bumper. The young Grenadier officer quickly located the Viking's company commander and was briefed on the plan. Hennessey was to drop down onto the canal bank and push forwards as quickly as possible using his vehicles in order to outflank the Taliban positions to the front of B Company. Hennessey returned to the eager Afghan company and briefed them on the plan. The ANA needed no extra encouragement to speed toward the enemy in their Ford Rangers. Although they lacked protection, the little pickup trucks were highly mobile and it didn't take long for Hennessey to lead the Afghan troops along the side of the canal. They were able to approach the enemy positions from the south once they dismounted and the Taliban now had no choice but to switch fire onto the advancing Afghan and OMLT soldiers. As soon as they got to about 500 metres from the Taliban-held compounds they came under effective fire. The OMLT WMIKs remained on the tow path and gave essential fire support to Hennessey's dismounted troops. The ANA had a 12.7mm Russian-made DShK heavy machine gun bolted into the rear of one of the pickup trucks and its heavy calibre bullets proved essential at this long range. The heavy rate of fire from the British and Afghan vehicles included the loud pop of the grenade machine gun as it delivered its lethal 40mm grenades onto the target. The vehicles were able to move up and down the tow path delivering a massive hail of bullets and explosives onto the Taliban. Before long the buildings from which the enemy were firing were engulfed in dust and smoke. Red tracer rounds disappeared into the smoke and occasionally shot skywards as they ricocheted off the buildings. Unlike their British counterparts, the dismounted ANA troops were lightly clad and moved quickly. They rapidly closed on the first compound which was entered and efficiently cleared. The Afghan soldiers were at their best advancing at speed through the mud-walled settlements and were well-drilled in this practice. Hennessey was impressed with the way they quickly moved forwards maintaining momentum. The British NCOs provided strident encouragement and leadership to keep things moving. The Grenadiers accompanied their ANA soldiers through the terrific noise and thick dust raised by the fighting. With fixed bayonets they used grenades to clear the little compounds and screamed instructions at the interpreters.

The Taliban, attacked from an unexpected direction, were now on the back foot and were unable to halt the steady ANA advance through the cluster of compounds. The enemy had been moving from one set of buildings to another and tried to delay the advance from small scrapes in the ground. After 1.5km the enemy ran out of compounds and tried to make a stand in the area of some sluice gates but with no buildings for cover were cut to pieces by accurate fire from the OMLT. The Afghans cleared through the area and found seven dead enemy fighters. Some of the wounded Taliban ran back into the arms of B Company who patched them up and handed them as POWs to the ANA who later delivered them safely into captivity in Gereshk.

B Company was now able to push up and secure the area. The plan had worked well, the Afghan troops in particular had carried out an impressive flanking movement. The company commander seemed pleased with the day's work and a joint clearance patrol was sent to push forward to the limit of exploitation, which was secured by nightfall. David was content with 1st Kandak. Not all of their officers were good but the fighting spirit of the soldiers was exceptional. They had responded well to British leadership and many lessons had been learned. The ANA had, however, taken three casualties. In addition to the soldier wounded to the south of the river, a sergeant lost an eye leading the assault and one of the platoon commanders managed to shoot himself in the foot while climbing through a window.

The Grenadiers had learned that the only way to get the Afghans to operate effectively was to lead by example. This meant that they had to be right up alongside the ANA soldiers in the assault. They needed to be led, encouraged and provided with tactical advice through their courageous interpreters who also moved alongside the assaulting troops. Hennessey was chuffed that his ANA troops had been able to carry out such an important assault to assist the hard-pressed B Company and that it had been the ANA who had been the first to reach the limit of exploitation. The weary troops settled down for the night close to the banks of the Helmand River. Operation Silicon was far from over and the Queen's Company would have another task in the morning.

For the OMLT, the next phase of the operation would be the establishment of three PBs which would be occupied by the ANA in order to provide a security screen in the area to the north of Gereshk. Together with B Company they had cleared the Taliban from the valley behind them and would now have to ensure that the enemy did not return to disrupt development in the towns and villages. David and his company HQ joined the rest of 1st Kandak in the morning. He would site and oversee the construction of the PBs. To the north, Operation Silicon continued with

A Company of 1 Royal Anglian forcing the enemy from the village of Habibollah Kalay. It wasn't long before The Queen's Company found that the enemy were not going to go away that easily. Mortar rounds started to impact on their side of the river and the Grenadiers searched frantically for the source of the deadly bombs before they got too close. The mortar team was spotted in the distance with the aid of binoculars, and CSM Simon Edgell's Javelin detachment launched a missile at the troublesome mortar. The lethal projectile whooshed into the air and descended onto its prey to devastating effect. The ANA and Grenadiers whooped with delight as a huge cloud of brown dust rose from the area of the impact. The Javelin was followed by some accurate British mortar fire for good measure. Unsurprisingly they were not troubled by mortars again that day.

The patrol bases were to be situated on the north side of the Helmand River. They were to run almost in a straight line south to north with the furthest being a little over 2km from the river. The southernmost base was to be on the banks of the waterway and would dominate one of the major crossing points. Before construction could be started, the area had to be cleared to ensure that there were no enemy fighters hiding in the nearby compounds. The ANA were enjoying the break from action and startled the OMLT by fishing in the nearby river with grenades. It was a dangerous practice and a waste of ammunition, but the Grenadiers soon learned it was impossible to stop the Afghan troops from participating in this activity which they regarded as being completely legitimate. The time was used for administration and it was necessary to remove a number of Taliban bodies from the area as they were starting to smell. Later in the day the engineers arrived to start construction on the PBs. Captain Walker and Staff Sergeant Blow organised their Afghan troops from the combat support kandak together with more British Sappers. It took 36 hours to build the three bases, which was a considerable achievement given the scant resources and incredible heat. They were extremely basic and consisted of little more than a bastion wall perimeter with some simple sangars situated for defence. Once they were completed they were occupied and the ANA settled in, together with their mentors. For the following two days there was little activity other than some sporadic shooting. It was assessed that at least 20 Taliban had been killed in the area during the operation.

On 7 May the Taliban delivered confirmation that they had no intention of staying away. A Royal Engineers Pinzgauer was caught in an explosion on one of the tracks to the south of the PBs. The device had been laid to ambush the British soldiers moving to and from the newly established outposts, but mercifully it had only partially detonated. The resultant explosion was sufficient to throw the vehicle from the road and to destroy it. The crew of the vehicle were able to walk away; it was an extremely close call and the beginning of a worrying trend. The Taliban were starting to realise that they would always lose in direct confrontation and were now resorting to laying mines and IEDs. There were also dangers other than the Taliban on the battlefield and this was starkly demonstrated when a British patrol from the Worcester and Sherwood Foresters Regiment (1 WFR) became engaged in a firefight with a joint OMLT and ANA patrol. A large amount of ammunition was expended by both sides and the situation was once again compounded by poor communications. Mortars were about to be called for when the mistake was recognised, the Grenadiers frantically tried to stop the ANA from returning fire and threw coloured smoke in the hope that the WFR troops would also realise their mistake. The firing eventually stopped and there were no military casualties, but it had been a close call.

Things were quieter in the area immediately north of Gereshk now. British troops took over some of the PBs and a routine was established. Worthington was now in command of the southern

PB and the Grenadiers there waited eagerly for news that they would be relieved some time soon. Eventually, David established the Joint District centre in Gereshk to the south and his company HQ moved there leaving Worthington and the platoon commanders to run the PBs to the north. Logistic support came mainly from FOB Price which was about 15 minutes away from Gereshk. Movement between the various locations was always tense and highly dangerous, the threat from mines and IEDs was incredibly high, so administrative movement was kept to the minimum.

The Afghan troops had proved their worth in battle but like most soldiers they quickly became bored with routine. On 9 May one of the young soldiers decided to teach himself how to drive. He clambered into one of the Ford Rangers and was able to put the vehicle in gear and to drive forwards. He steered the vehicle out of the PB and straight into the fast-flowing Helmand River. The OMLT troops, alerted by the excited Afghan shouts, arrived in time to see the ANA truck bobbing up and down, moving rapidly downstream with its driver still at the wheel. The ANA set off in pursuit of the fast-moving vehicle which fortunately lodged against a bridge some distance down the river. The truck was recovered with help from the Grenadiers and the bedraggled driver seemed no worse for the experience. The offending soldier was duly returned to the joint distric command centre where he was locked up awaiting a decision on his punishment from the ANA commanding officer.

Operation Silicon formally ended on 14 May but the Queen's Company troops were told that they were to remain in place indefinitely until a relief could be arranged. The whole of Task Force Helmand was involved in offensive operations and this was not a good time to organise a relief. There were continual small-scale skirmishes with the enemy. On 17 May the company pushed 2km forward from the PB to provide flank support for a thrust by B Company of 1 WFR into the Green Zone around Rahim Kalay.

As B Company pushed into the Green Zone, a group of Taliban were spotted by the Grenadiers. Worthington called in mortars but these were delayed because the airspace was occupied. One of the company snipers engaged the enemy at long range and his target was seen to fall. The range was estimated at 1,500 metres, an incredibly good shot by anyone's standard. That evening an ANA patrol struck an IED and two soldiers were seriously injured, one losing an arm. Three days later ten bodies were found about 3km from the PBs. They turned out to be Afghan National Police who had been captured by the Taliban and ruthlessly executed. These stressful daily events continued as the summer heat became increasingly unbearable for the troops dispersed in the growing number of isolated PBs around Helmand.

10

SECURING SANGIN

The Royal Engineers had completed their work on the Sangin patrol bases some weeks earlier. Poppy fields had been flattened and tons of sand had been dumped into the bastion walls. The barriers provided some protection for the troops who occupied these dangerous outposts. For the most part the PBs were built on abandoned compounds and the existing structures were fortified by whatever means the engineers could manufacture. Barbed wire, sangars and a few claymore mines were added. Most of the locations had sandbagged positions on the flat roofs from where the defenders could observe the surrounding area. US troops had been occupying the new locations together with some Afghan reinforcements from Kandahar, but this was a temporary arrangement. Once Operation Silicon was completed, UK OMLT and Afghan troops from 3/205 Brigade were to relieve the Americans. Although Major Martin David had been so heavily involved in the siting process, neither he nor the Queen's Company were destined to occupy the bases just yet. They were now garrisoning the Gereshk PBs and the joint district command centre together with 1st Kandak. Lieutenant Colonel Carew Hatherley and General Gori Muhayadin had decided to leave them in place and to move 2nd Kandak with their 2 Company mentors up to Sangin.

In early May, 2 Company, with their Afghan colleagues, relieved the Americans. Each PB location was to be manned by an ANA platoon from 2nd Kandak. The 2 Company mentors would be split down to support the deployed Afghan soldiers. This meant that each PB would be manned by four or six Grenadiers and between 30 and 40 Afghans. The kandak HQ would remain in FOB Robinson so the OMLT mentors were now very thinly spread. The OMLT and ANA were responsible for patrolling the area of the town and for demonstrating an Afghan presence. A British infantry company was stationed at the district centre but they operated largely independently. Sangin was an incredibly dangerous place; the Taliban had been kicked out but many of them had remained in the area covertly or were creeping back into the locality to harass the coalition troops. Route 611 was now a major target for them. The route had to be used through the town and it was easy to lay mines under the cover of darkness. The PBs, being more isolated, provided easy targets and the Taliban attacked them right from the very start.

The three new defended locations were initially referred to by their geographic location, but this was to cause some confusion as the Gereshk PBs were now similarly named. The Sangin bases were later named from north to south as 'Blenheim', 'Tangier' and 'Waterloo' after Grenadier Guards' battle honours. Each location had its own characteristics and vulnerabilities.

The northernmost PB was Blenheim. The compound was a couple of kilometres from the District Centre and sat astride Route 611 as it left the town and headed north to Kajaki. The high compound walls provided some security but the entrance and extended area were very exposed. This area consisted of a single bastion wall about four feet high and some barbed wire. A small sangar was home to two Afghan sentries who overlooked the Green Zone and the steep ramp into the PB. Each of these defended locations was similar in construction, although the buildings and views over the ground varied.

Staff Sergeant Dave Harrison of the Royal Artillery was to take over Waterloo PB; he had only three Grenadiers to mentor an entire platoon of Afghans. Before departure from FOB Rob, this small group of men were alarmed to hear that an IED had exploded just outside the entrance to their new base. There were US and Afghan casualties who were now on their way to FOB Rob. Harrison was directed to move off regardless and take over from the US troops as planned. Harrison was accompanied by Lance Corporal Shread, Guardsman Lawa and Guardsman Elasi. Together with their ANA charges they made the short 15-minute journey to PB Waterloo. When they arrived they found the Afghan garrison mounted in their rangers, engines running, eagerly waiting to depart. Harrison was surprised to find no sign of the US mentors; when he questioned the Afghans he was told that the Americans had already left. There was clearly going to be no formal handover.

Together with the three Grenadiers, the NCO quickly toured the PB to get a feel for it and to make an assessment of what needed to be done before the outgoing Afghans departed. He was dismayed to hear the roar of engines and to suddenly see the overloaded rangers driving away from the PB without a backward glance. Harrison rushed to his Afghan platoon commander and was horrified to hear that his ANA detachment were quite content with the handover and had said farewell to their predecessors just 15 minutes after they had arrived. With some urgency Harrison directed his three Grenadiers to organise the defence of the 100 x 100 metre perimeter. Sentries were posted in the four empty sangars and the first initial crisis was overcome. There was no barricade across the entrance to the PB and Harrison decided something would need to be put in place there urgently. There was a small detached observation post just the other side of Route 611 which also had to be manned. The little group of British soldiers assessed that their outpost was at this stage still very vulnerable to enemy attack. The tiny OMLT detachment was then further alarmed to find that it was unable to establish communications with

FOB Rob or with anyone else for that matter. Harrison quickly discovered that his radio had completely given up. To remain in the exposed location with no communications was simply not an option. The NCO decided to take one other Grenadier and an escort of ANA soldiers and to make the dash back to FOB Rob to exchange the broken radio. It was an uncomfortable decision for Harrison to make but the party reached their destination safely and then returned to their isolated PB in the early evening with the new radio. The four British soldiers now mounted a patrol of the perimeter, two men on the outside and two within the scant defences. A few vulnerable areas were found – ditches that had not previously been seen and a few caves – all of which would provide cover for the enemy. Harrison organised the defences accordingly and the occupiers of PB Waterloo settled into a night routine.

At around 2030 hours an excited interpreter came running up to Harrison to say that he had received some dramatic local intelligence. There were apparently 50 enemy fighters planning to attack the west side of the PB, they had dug a trench and were in the final stages of preparation for the assault which was to be initiated by RPG. The four British soldiers quickly rallied the ANA platoon and positioned them around the perimeter. Each of the four mentors had about eight Afghans and, once in position, they stared into the darkness waiting for the storm to commence. The interpreter relayed information to the troops, becoming more alarmed as he did so. Before long the distressed man was near to breakdown and was telling Shread that 'they would all die here tonight'. Shread played the comments down, but was secretly more concerned than he was willing to admit. Harrison had been trying to reach FOB Rob on the radio but, disturbingly, he was again unable raise anyone. The tension continued to rise as the interpreter informed Harrison that the enemy were in the trench and ready to go; everyone waited for the attack. The Afghan then exclaimed that the enemy had reported that 'tanks' were approaching and to hold the attack. There were no tanks in the Sangin area but the troops knew well that the enemy often described WMIKs as tanks. After straining their eyesight and hearing, the besieged party were able to establish that a small convoy was indeed moving down Route 611 from the direction of Sangin district centre. Harrison ran out into the road, relieved that the British numbers had just increased substantially; he found the company commander and hurriedly explained the situation to him. The reinforcements quickly drove into the PB and took up defensive positions. The slightly more composed interpreter was now able to report that the enemy had postponed their attack. A tense first night was spent in the little outpost and everyone was reassured when daylight finally came. A dawn patrol revealed a freshly dug crawl trench only 20 metres from the west wall of the base. It was about two feet wide by two feet deep and had clearly been occupied only a few hours before. Harrison realised how fortunate they had been and how timely the arrival of company HQ was. Four British soldiers with a small platoon of ANA soldiers would have had a serious fight on their hands.

One of the key tasks for the OMLT detachment in PB Waterloo was to clear the route from Waterloo to Tangier base daily. No one else was supposed to use the route until Harrison had reported it clear. This was a dangerous and nerve-wracking job. The enemy were easily able to infiltrate and place IEDs on the road under the cover of darkness and this section of Route 611 soon become known as 'IED Alley'. Having only just recovered from the tense experience of the previous evening, Harrison's group now prepared themselves for the morning's dangerous task. As they went about their preparation they were surprised to see two ANA vehicles approaching at speed along Route 611 from the direction of FOB Rob. No one was as yet cleared to use the route for the day but the ANA were operating independently without clearance. Despite constant warnings and pleas from the

British and American mentors, it was common for the Afghans to drive into the bazaar to purchase rations.

Harrison's patrol set off along the route, carefully searching for the tell-tale signs of roadside bombs. There was a sudden huge explosion as dust and dirt was thrown into the air, and the British troops instinctively ducked for the cover as shrapnel fell to earth around them. An IED had detonated at the side of the road only 100 metres ahead. The ANA vehicles had both been carried through the explosion by their speed. Remarkably the occupants were unharmed and the vehicles just kept going, their passengers whooping with delight as they sped past the alarmed British patrol. The two pickup trucks careered down the road leaving a dust trail behind them as if nothing had happened. Harrison's men now had to clear the site of the explosion and the remainder of the route; to say they were not amused would be an understatement. After cautious exploration, Lawa located the firing point for a command wire device. It had been fired from only 50 metres and the firing pack was still in place. The remains of an old 105mm shell were also discovered. The device had been concealed inside a tyre at the side of the road and was only 25 metres from the device that had blown up the Americans the day before. It occurred to the British troops that, had the IED not been initiated against the fast-moving Afghan trucks, it could well have taken out Harrison's patrol.

The following days and weeks fell into a similar pattern and the occupants of the PBs came into contact with the enemy on many occasions. Route 611 lived up to its reputation as Harrison's party narrowly avoided a whole series of IED attacks. Sometimes the devices were located or went off prematurely and the tired troops rode their luck as they waited to be relieved. Lawa always said a Fijian prayer before the patrols and the others hoped that the prayers would continue to work. It was a nerve-shattering business which took its toll on weary men. On 17 May a large convoy of troops from the Royal Anglians stopped at PB Waterloo. They were en route for Sangin and were keen to see the ground from the PBs. Harrison provided a ground brief to the company commander, warning him about the threat of IED attack and ambush in the area. He pointed out the now notorious blue building which was often used as an enemy observation point. The Vikings joked with the Grenadiers and asked what they had done wrong to be abandoned in such an isolated location. After a short period the convoy moved off once again, leaving the occupants of the PB feeling rather alone. A short time later the line of British vehicles was ambushed. Some 500 metres south of PB Tangier a large force of Taliban fighters fired machine guns and RPGs at the convoy. A Viking tracked personnel carrier was hit by the deadly rockets and was destroyed; there were a number of casualties. A fierce firefight ensued and Apache helicopters overhead were able to kill a good number of retreating enemy fighters. The fight was for the most part out of range for those in PB Waterloo, but Harrison was able to direct the fire of the helicopters. As the battle died down, thick smoke could still be seen coming from the now burning Viking personnel carrier. The wrecked armoured vehicle would return to haunt the Grenadiers in the days ahead.

Martin David was proved correct over the siting of Tangier base. The blue house that they had wanted to use, but had been so significant to the ANA during the negotiations, was repeatedly occupied and used as a firing point by the enemy. It seemed that the Taliban agreed with David over the commanding position of the house and they used it frequently to launch their attacks on both Waterloo and Tangier. Ironically, the firing point became such a thorn in their side that Major Toby Barnes-Taylor decided that access to the upper levels should be denied by blowing the stairs. In a true comedy moment the engineers placed explosives where they would destroy only the stairs. When the charge was detonated, in fact, the whole house came crashing down in a cloud of white smoke. When it cleared nothing but a pile of blue and grey rubble remained. Some wag was heard to say in a Michael Caine-like voice, 'I only said blow the bloody doors off!'

2 Company patrolled the area daily and joined US troops to flush out the Taliban from areas of the Green Zone so that they could be targeted. There were several contacts with the enemy, but the Taliban always came off worst. The OMLT operation was controlled from FOB Robinson and the company HQ patrolled out to the isolated PBs frequently. The ANA were introduced to night ambush operations in an attempt to deter enemy activity. This was a new concept to the ANA but confidence was built when they found that they were capable of mounting the ambushes with British leadership. At FOB Rob, CSM Darren Westlake guided the kandak HQ through the finer points of administration. While checking the ANA building he discovered a cellar with a small number of Taliban prisoners confined within it. Westlake was concerned that the enemy fighters might not have been treated to the high standards required of the British Army and he provided some strident education to the Afghans on prisoner handling. He also arranged for the prisoners to be put in British custody with the Military Police. The Afghans were learning, but it was a slow process.

At PB Tangier, Captain James Shaw took command and struck up a good relationship with the Afghan platoon based there. The ANA gathered a great deal of local information and intelligence which was passed through their own reporting chain. Not all of the information was translated and passed to the British and there were sometimes surprises. On 11 May, Shaw was taken aback to see his Afghan charges mounting up and preparing to leave the base. After some urgent questioning it was discovered that the ANA had gathered some 'good' information on the location of a Taliban weapons cache. Shaw urged caution and tried to explain the potential perils of racing off without proper consideration. The ANA commander would have none of it and insisted on leaving immediately. The OMLT troops had no option other than to follow.

The ANA trucks raced through the centre of Sangin and then headed north. Before they knew it Shaw and his small party were passing PB Blenheim and were still heading north on Route 611. They were now in very dangerous territory and Shaw urged the Afghans to stop, but his warnings were not heeded. His worst fears were realised when the convoy came suddenly under intense small arms and RPG fire. The British troops immediately returned fire and slammed their vehicles into reverse to extract from the killing zone. The whole thing had clearly been a ruse and the ANA had taken the bait. The OMLT drivers managed to get the vehicles away from the worst of the fire and Shaw now took stock of the situation. There were no British casualties, but the ANA had dismounted in the middle of the Taliban killing area and were now pinned down in a ditch. Shaw realised that the Afghans would probably stay there until the enemy decided to flank them, so drastic action was needed. Together with Sergeant Byrne, he dropped into the ditch and the two Grenadiers used the little depression to cover the 200 metres or so to the isolated Afghans. As they approached, the crack of high velocity rounds grew louder and bullets thumped into the earth nearby. When they reached the stranded ANA troops it was clear to see that they were in a state of confusion and were receiving little or no leadership from their own commander. Three of the ANA soldiers had been shot and were in a serious condition. One of the vehicles had been struck by an RPG and smoke hung over the entire area.

Shaw shouted to the Afghan commander that they had to withdraw and used hand gestures to make the Afghan understand. Byrne moved along, conveying the message to the young soldiers as bullets cracked overhead. It took a while but eventually the message was understood and the ANA troops started to follow, keeping as low as they could. Covering fire was provided by Sergeant Brooks, Guardsman Humphries and Signaller Busby, while Lance Sergeant Robinson radioed in for a 105mm fire mission from the FOB Rob battery. The guns would soon deliver accurate shells onto the enemy, but the deadly explosives would be dangerously close to Shaw, Byrne and the withdrawing ANA troops. As he led the way out of the killing area, Byrne looked back and to his horror saw the ANA drivers clambering into the cabs of their vehicles in an attempt to recover them. He screamed at them to leave the vehicles as bullets kicked up earth and dust all around the Ford Rangers. Incredibly, the Afghans were now performing three point turns on the road in the middle of an ambush. He was amazed to see the vehicles race past him on the road heading south to safety, being pursued by RPG rockets and small arms fire. It was a miracle that the drivers had not been killed.

The line of ANA soldiers followed Shaw and Byrne who were also under RPG fire. Shooting was now coming from a different direction, indicating that the enemy had tried to flank the ANA party. Were it not for Shaw and Byrne the escape route would have been closed and most of the ANA men would have been killed. The desperately needed fire mission never arrived; it was once again delayed due to the presence of coalition aircraft in the area. Company HQ, together with a patrol led by Captain Ed Janvrin, who had responded to the alerts from Shaw, now arrived on the scene from FOB Rob and provided cover for the withdrawal. It was a slow process, but eventually the Afghans were led out of the ambush without taking further casualties. The wounded ANA soldiers were placed onto the British vehicles and the whole party withdrew to FOB Blenheim where the Afghan casualties were properly treated before moving into the Sangin district centre for extraction by Chinook.

When they arrived back in FOB Tangier, Shaw and his party were exhausted but there was to be no rest. That night the Taliban attacked Tangier and a protracted firefight took place as the enemy attempted to probe the sentry positions and the nearby vehicle checkpoint, which was known as 'VCP Suffolk'. The enemy broke off contact after a lengthy gunfight and did not appear again that evening. It was none the less a nervous night for the handful of OMLT troops who were by now very conscious of their isolation.

The vulnerability of the PBs was demonstrated once again on the night of 15 May when at 1935 hours PB Tangiers and VCP Suffolk were attacked in a determined and coordinated Taliban assault. Shaw and his small OMLT team were sitting and chatting through recent events when the base exploded with machine gun and RPG fire. The Grenadiers reacted immediately. Byrne and Robinson returned fire with GPMGs and Brooks rapidly launched bombs from the 51mm mortar. Shaw tried to call for assistance only to find that the radios had suddenly decided to give up. The Grenadiers estimated the enemy to be around 30 strong. The gunfire could be seen and heard from the nearby Sangin district centre but no-one was sure of exactly what was happening.

The situation deteriorated further when the Afghan commander in PB Tangier reported that VCP Suffolk had been overrun by the Taliban. In Sangin district centre, approximately 1km away, A Company of 1 WFR reported seeing lines of tracer streaking into the PB and the flash of RPGs as they exploded. They had two platoons and their mortar line ready to assist, but could not establish communications. They could hear Shaw asking for support but could not reply. In FOB Rob, 2 Company HQ could also hear the request for assistance and Barnes-Taylor and Westlake immediately set off in three WMIKs while the US troops in FOB Rob called for Apache and other air support. The ANA and OMLT continued to return fire and the Taliban attack eventually lost momentum. In fact, VCP Suffolk had not been overrun. It had held, thanks to the incredible bravery and determination of the ANA defenders who had fought a close quarter battle in a lightly defended and isolated position against a numerically superior enemy. For the second time in days, the Taliban had showed themselves to be a formidable foe capable of mounting well-planned and coordinated attacks. The Grenadiers and their ANA colleagues had done well but they were now concerned about their vulnerability and they wondered what the enemy would do next. During the following two evenings, VCP Suffolk was attacked again and Shaw and his men were obliged once more to give support to the besieged ANA troops. Sangin was certainly living up to its reputation and the Grenadiers wondered how long the Taliban could maintain such a tempo of attacks against them. During the action on 11 May a Canadian journalist was present with Shaw and his troops. He wrote a front-page article for the Canadian *Globe and Mail*. Days later Shaw's cousin was amazed to read of his exploits on the front page of Canada's most prominent newspaper. The cousin thoughtfully telephoned Shaw's father to inform him that his son had been in a 'spot of bother'.

On 21 May the Taliban changed their tactics and the war in Sangin began a new and dirty chapter. The day's events would clearly demonstrate the ruthlessness of the enemy and their total disregard for the civilian population. The burnt-out carcass of the Viking armoured vehicle destroyed on 17 May still lay in the middle of the road as a reminder of Taliban capability. The OMLT troops suggested that it should be removed as it did nothing for the confidence of the locals. There were further concerns as it formed a vulnerable point on the road where an IED could be placed. These worries went unheeded as there were other priorities at the time. Unfortunately, OMLT fears were confirmed when at 1235 hours a loud explosion was heard from the direction of the wreck. Investigation soon revealed that the Taliban had placed a device inside the hulk. Three local children who had been playing on the wrecked vehicle had disturbed the bomb, causing it to explode. Bryne and his party were first on the scene. They were horrified at what they found. One child had been blown apart leaving just a bloody mess, the second was also clearly dead and the third child had lost his leg and was dying. As Byrne knelt to treat the child he saw the desperation in the young boy's eyes; it was clear that he was minutes from death. The wounded child was rushed to the Sangin district centre where there was a doctor, but died shortly after. For those involved this was a most disturbing experience and would live with them for a very long time. The Grenadiers had received training in the management of psychological trauma and in body handling, but nothing could prepare them for an experience such as this. Everyone was angry at the Taliban who had shown no regard for the safety of the civilian population. No coalition troops were near the wreck when the bomb exploded and the deaths were just so unnecessary.

The protection of Afghan civilians was at the core of ISAF military strategy and the troops worked extremely hard to avoid civilian casualties. However, the local population were caught in the middle of a desperate and dirty war and on occasion became the victims. Only a week earlier, another incident had occurred when US troops had called in air support during a heavy contact. Bombs were dropped and the Taliban as usual broke contact. The following day many wounded civilians were brought into FOB Rob, where the limited medical resources were overwhelmed. CSM Hill worked hard to assist with the first aid in the absence of sufficient qualified medics. Medical teams were flown in from Bastion to assist and many of the wounded were evacuated by Chinook. During the afternoon a group of civilians arrived at FOB Blenheim where they revealed the bodies of more dead children. They claimed that they had been killed during the fighting the previous evening. These were dreadful and frustrating sights for young men to endure but they were powerless to do anything except report the facts up the chain of command for investigation and to assist the wounded. The Taliban exploited the situation at every opportunity; they sheltered among the civilian population knowing full well that they put them at risk. The civilians were often forced to carry ammunition and weapons. Only the previous week, a US unit had seen women and children passing ammunition to the fighters who had ambushed Shaw's party. Coalition troops were restricted by their rules of engagement and steps were always taken to positively identify the target. The enemy, who caused the vast majority of civilian casualties and targeted them indiscriminately, took no such steps.

The effect of large numbers of British, US and Afghan troops in Sangin was marked. Before long, local traders returned to the bazaar and despite the fact that fighting was still going on, some form of normality started to return to the town. Half of the buildings were in ruins but there were plenty of people still living there and they needed to buy food and goods from the shops in the town. The task of the ANA was to give people the confidence to continue to do so and to keep the Taliban outside the populated area.

2nd Kandak was overdue for leave and would at some point need to be replaced by 3rd Kandak who had recently returned. This meant that the Inkerman Company would replace 2 Company mentors in the Sangin area. At the end of April, Major Marcus Elliot-Square and a small group from the Ribs moved up to FOB Robinson to conduct a reconnaissance of the locations that they would inherit from 2 Company. The long journey took more than six hours and was incredibly tiring. There were precious few comforts to be had in FOB Rob when the recce party arrived and it was clear that enemy rocket attacks were still causing problems for those who were stationed there.

Elliot-Square spent the following days looking around the various PBs and areas that he would be taking over. Janvrin was a great source of information; he had spent longer than anyone in the Sangin area and had recently seen action as the British liaison officer to US troops who were conducting mobile operations around the district. Their mission complete, the Inkerman recce party departed for Shorabak. It was necessary to join a larger logistic convoy that was heading south and the journey was plagued by a series of setbacks. Vehicle breakdowns, bogged in trucks and other problems meant that a six-hour journey turned into a 17-hour nightmare, but Elliot-Square and his little group eventually arrived back at FOB Tombstone exhausted but in good order. The following days were occupied with detailed preparations for the company and kandak move to their new area of operations. In addition to occupying the three newly established PBs in Sangin, elements of the company would be taking over in Lashkar Gah and in Kajaki. The company HQ would remain in FOB Rob; like 2 Company, the Ribs would be widely dispersed.

The Inkerman Company finally left Camp Shorabak for Sangin on 25 May. They had been scheduled to depart much earlier than this but Afghan time-keeping as usual slowed things down. The convoy consisted of UK and Afghan vehicles full of the men and equipment needed for the lengthy stay in the outstations to the north. The long snake of vehicles moved slowly north onto highway 1 and then east through Gereshk. There was a significant threat from suicide bombers in the town and everyone was on their guard looking for the warning signs. Once through the busy little town, the convoy eventually turned north into the desert for the long cross-country drive to FOB Robinson and Sangin. Although the convoy now had to slow down to allow the heavier vehicles to negotiate the rough desert terrain, this would be a much safer route than the dangerous Route 611. The convoy was not safe from Taliban attack, though; an established enemy tactic was to use mopeds to monitor the progress of the convoys. Once the direction of the convoys had been established the riders would speed ahead to lay mines in any potential wadi crossing points or bottlenecks. As this particular convoy was more than 2km long and was producing a huge dust cloud visible for miles, everyone searched the horizon for any sign of motorcyclists. The big American 6 x 6 trucks driven by the ANA were very slow and lumbering; their off-road capability was poor and the convoy was continually forced to halt as they caught up. The baking heat made it incredibly difficult to concentrate and the drivers struggled to keep their wheels in the vehicle tracks ahead of them. Under the desert sun the plastic bottles of drinking water soon became hot and the clear liquid contained within became almost impossible to drink. A useful trick quickly learned by the troops was to soak an army sock in water and to then slide the sock over the bottle like a sleeve. This helped to slow the heating process and to keep the water cool. Elliot-Square continually altered the axis of advance in order to confuse the Taliban who would almost certainly be monitoring the convoy's progress. He was, however, frustrated at the slow rate of advance and the continual vehicle breakdowns. The REME mechanics accompanying the patrol worked incredibly hard under the desert sun to keep the convoy rolling and Captain Dave Groom, the transport officer, had once again come along to lend his assistance.

Some time during the late afternoon Elliot-Square was notified that one of the old American trucks was again bogged in. He halted the convoy and patiently waited for the truck to be extracted from the deep sand. After a long period with no news of the vehicle being freed, he decided to investigate. The Grenadier officer was horrified at what he found. Several of the ANA heavy trucks had become stuck fast in the soft sand. This was clearly a situation that was not going to be resolved any time soon, the light was fading fast and it was obvious that FOB Rob would not be reached before dark. Elliot-Square decided to spend the night where they were and the Afghan battalion was arranged in a defensive perimeter. Just when the Grenadiers thought that things couldn't get much worse, a huge sandstorm blew in and an uncomfortable night was spent by the soldiers who tried to keep the thick sandy dust from their eyes and mouths.

At first light the sandstorm had died down and efforts to extract the Afghan trucks were renewed. It soon became clear that it was not going to be possible to drag the vehicles out of the soft sand. They did manage to save one of the trucks, but Elliot-Square had little choice other than to order that the remaining vehicles be 'denied' to the enemy. The ANA swarmed over the trucks quickly stripping anything that could be salvaged, including seats and even the doors. Plastic explosives were attached to the engine blocks in order to render them useless and red phosphorous grenades were used to torch what was left. The Inkerman Company then drove away from the scene leaving the burning hulks in their wake. The convoy arrived at FOB Rob later that day and the Inkerman men were able to swap stories with some of their 2 Company comrades, most of whom they had not seen for a very long time. Meanwhile, Captains Phil Hermon and Vince Gilmour (two attached officers from the Royal Tank Regiment and Royal Regiment of Scotland respectively) had arrived 24 hours earlier with a small advance party. Over the following day the Ribs were to relieve the tired men from 2 Company who would guide their 2nd Kandak soldiers back to Shorabak.

11

HOLDING THE LINE

For the men of 3 Company in Garmsir, May had been a tough month and the loss of Guardsman Simon Davison in the first week was still keenly felt. The Taliban had maintained the frequency of their attacks and had continued to try and get as close as they could to both the eastern checkpoint and JTAC Hill. The Grenadiers and the other British troops in FOB Delhi were now acclimatised to the intense heat and the isolation of the location. Visitors to the outpost were often quite shocked at the appearance of the men there. Like their predecessors they had already lost weight and many wore beards. There was a sense of 'being out on a limb', of being on the front line, and the troops took a perverse pride in their situation. Everyone complained about the lack of creature comforts but few would have voluntarily left the company for a more cushy existence further north. Coalition aircraft were usually on station overhead and an air strike could usually be called in fairly quickly. The enemy positions were well known to the Grenadiers in FOB Delhi and the civilian population to the south had long since left the area. Any movements in the ruins and undergrowth to the south were certainly enemy forces and there was very little risk of civilian casualties. The 81mm mortars in the FOB fired daily and the crews had become slick and efficient.

The 105mm guns at FOB Dwyer regularly fired too and the Grenadiers usually watched the artillery display as it wrought death and destruction upon their Taliban attackers. Major Tim Law's CS

Kandak provided a number of Royal Artillery mentors to the ANA gunners at FOB Dwyer. There were some ancient Russian D30 guns located at the FOB and the OMLT troops tried extremely hard to train the Afghan gunners to become proficient in their use

The young Guardsmen had become very capable in the use of their heavy weapons and hundreds of GMG rounds had been delivered onto the enemy. The determination and fighting spirit that had developed after Davison's death had not faded. Night fighting patrols were now the norm and were conducted frequently. The enemy to the south and east of the town were never quite sure where they might encounter the British after dark. These fighting patrols took a great deal of preparation and not a little courage to step out into enemy territory in the darkness. British night vision devices provided a vital advantage over their concealed enemy, but a Taliban fighter concealed in a trench or tunnel would be virtually impossible to spot in the dark. The patrols were consequently always tense affairs. The targets for these missions were generally areas that were known to be occupied by the enemy or where there had been recent activity. Denying these positions to the cunning and ruthless fighters who sought to conceal themselves there would prevent them from establishing fortified areas closer to the town from which they could fire with impunity. The fighting patrols were thus intended to force back the forward line of enemy troops.

On 25 May, 3 Platoon, under the command of Lieutenant Andrew Tiernan, prepared for another of these fighting patrols. The Guardsmen were accustomed to these operations and accepted them as being fundamental to their mission in Garmsir. Tiernan delivered his orders to the troops in good time for them to carry out the battle procedure necessary to mount the patrol effectively. 3 Platoon was to patrol south of the area of their objective, which was a series of old farm buildings that were known to be occupied by the enemy. Two days earlier Tiernan had led a reconnaissance patrol to the south. The patrol was designed to try and identify the enemy forward positions. It was believed that the enemy held a defensive line along an old ditch known as 'Route Taunton' and it was hoped that this could be confirmed. The patrol moved south using the Helmand River as a shoulder. Tiernan managed to get within 30 metres of the Taliban who were clearly heard talking on their radios. The patrol withdrew but not before positioning itself to look along the enemy front line from almost right angles. It was a very successful endeavour which produced some useful intelligence for the fighting patrol that was now preparing to leave.

The troops occupying the JTAC Hill observation post would provide overwatch of the area with their night vision devices. The platoon would move silently on foot through a series of ditches, hoping to take the enemy by surprise. A fire support group with additional GPMGs and a Javelin detachment would provide extra firepower from the west near the river bank. The company snipers were to be positioned to the east and would warn of any enemy activity from that direction. Once everyone was in position the attack was to be initiated by firing a Javelin missile into the area of a suspected sentry position. The Grenadiers would then clear through an identified trench system destroying any enemy encountered; a reconnaissance of a hitherto unknown network of trenches would then be conducted before returning to FOB Delhi.

The patrol crept silently from FOB Delhi at around 2300 hours. The darkness in Helmand falls early and covers the countryside in a black veil that aids those who choose to move covertly. There was little light other than the dim reflection of the moon on the Helmand River, which the platoon soon left behind them as they dropped into the ditch that would lead them south towards their objective. The Grenadiers were travelling as light as they could. All non-essential equipment had been left behind, but the weight of their body armour, weapons and ammunition alone made the going tough. The lead scouts moved cautiously along the gully and stopped frequently to scan the area with the infra-red monocles that were fixed to the front of their helmets by hinges. Each man attempted to make out the outline of various reference points in the darkness, but it was difficult to see much. The going was slow and the troops were hindered by the number of large bomb craters in their path which increased as they moved further south. Bayonets were fixed as it was quite possible that a close quarter encounter with the opposition could occur. A reference point known as 'Frog Eyes' was cleared en route; this area was assessed as an enemy cache for ammunition and rations but on this occasion it was empty.

The Fire Support Group had split from the main patrol and now moved to its position along the banks of the river to the west. It took a little over an hour for Tiernan to lead his assault group to within striking distance of the objective and things had so far gone smoothly. Back in FOB Delhi the mortar men waited for Tiernan's signal to commence a bombardment onto the known enemy positions to the south; this would prevent the Taliban there from interfering with the assault. The young platoon commander now pushed one of his three sections around to the left. They would occupy a ruined farm complex and would be able to deliver fire onto the enemy position from a better angle, which would allow the assaulting troops to break into the first compound. The plan was textbook and had so far been flawlessly executed by the well-drilled British troops. As Tiernan confirmed that each of the groups was in position, he was acutely aware that the most dangerous part of the operation was still to come. Movement had been detected on the objective and the enemy sentries were obviously awake.

When he was satisfied that everyone was ready, Tiernan gave the order for the Javelin missile to be fired to initiate the attack. Guardsman Elliott Hennell had for some time been observing the enemy sentry position, waiting for the word to launch his deadly projectile. Lance Sergeant Pancott, in charge of the Javelin detachment, relayed the order to fire, but Hennell hesitated. The young Guardsman had suddenly picked up a second enemy heat source heading towards the original target. He realised that the sentries were preparing to change over and decided to take them both out. He waited for what seemed an age before pressing the firing switch and launching the missile. The silence was shattered by a deafening explosion; there was a flash of light and sparks flew into the air around the site of the impact. Hennell had totally destroyed the sentries and their relief. Simultaneously the section deployed on the left opened up and the air was filled with the sound of small arms fire. Orange streaks of 7.62mm tracer shot across the landscape as the Fire Support Group joined in. As Sergeant Scott Roughley led his assault section into the first compound the rapid crump of mortar rounds could be heard to the south as they fell onto the Taliban positions there. Those who watched the spectacle from a distance were able to make out the sound of hand grenades exploding inside the ruined buildings as well as rapid bursts of machine gun fire as Roughley and his men pushed through the compounds. The assault sections lobbed grenades into any likely hiding place and followed this up with quick bursts of fire into the blackened spaces. Roughley's men made good progress through the first compound and moved rapidly through the second. There was no sign of the enemy in either location; they had likely made themselves scarce as the assault moved towards them.

The next phase of the plan required the third, as yet uncommitted, section to attack the final part of the objective, while the first assault section gave covering fire. Roughley's men now took up firing positions along a prominent earth mound. This would be a good position from which to support the final phase of the raid. The air was still alive with tracer and the smell of cordite as the last section, led by Corporal Paul Morgan, passed through Roughley's panting men and took up positions to his left along the same embankment. The final section was about to launch its attack when a blinding flash of light and a huge shockwave tore through everyone in the area. A massive explosion threw earth into the air and Roughley found himself blown off his feet and dumped into the bottom of a water filled bomb crater. Dust and sand shrouded the area as it showered back down to earth and there was momentary confusion as everyone tried to gather their wits and make sense of what had just happened. Tiernan now moved forwards to the seat of the explosion in order to assess what had happened and how many casualties there were. A sudden piercing scream brought confirmation that there were indeed injured Grenadiers. Tiernan tumbled into the bomb crater where Roughley was now regaining his senses. The Taliban had started to return accurate fire and seemed to be homing in on the source of the terrible screams coming from the scene of the detonation. It was uncertain what had caused the explosion but the most likely explanation was that the Taliban had mined the approaches to their position, perhaps even remotely detonating the device.

Roughley crawled to the source of the screaming where he found 20-year-old Guardsman Scott Blaney writhing in agony. The young man was obviously seriously wounded and Roughley now gently told him to stop screaming; tracer rounds were zipping overhead and the section commander realised that the enemy were firing in the direction of the noise. Blaney gritted his teeth and did as he was instructed by the young NCO. Roughley now turned his attention to providing urgent first aid; Blaney's right leg had been shattered by the blast and he was losing a great deal of blood. He would bleed to death unless a tourniquet was fitted, and fast. Morphine was administered to help with the obvious pain. Blaney's right arm and the side of his face were also badly damaged and he courageously concentrated on not betraying his platoon's position to the enemy, who now held the initiative. Tiernan called for Morgan to clarify what had happened. Morgan was unable to reply at this stage because he too had been injured. Shrapnel from the blast had peppered his face and flash burns had blinded him, he also had a badly broken leg, although no one knew this yet. The situation was chaotic and confusing but there was even worse to come. There was an alarmed shout as Guardsman Langridge reported that there was another casualty near him. Roughley moved along the crater to find Lance Corporal Nick Davis who looked to have lost his leg and had severe injuries to his buttocks. Remarkably, he sat up and started to talk to Roughley, Davis was obviously in shock and needed urgent medical attention.

Sergeant Rodney Clarke tried to coordinate the first aid to the casualties while at the same time working out how the seriously injured men could be evacuated. Tiernan was now appraising the situation; he needed a new plan, and quickly. As he moved around, he suddenly came across a prostrate form. At first he thought it was a dead Taliban fighter, but as he looked closer in the dim light he could see that the body wore British uniform. The body was that of Guardsman Daniel Probyn, who had been killed instantly. Morgan's section had been almost entirely taken out by the huge blast. It was clear that the platoon would have to withdraw quickly if the lives of the seriously injured men were to be saved. As he made his new plan, Tiernan's interpreter reported that the Taliban were now attempting to surround the platoon. There was no time to lose.

Tiernan announced that the platoon were to move back to the west where they would meet up with the Fire Support Group. He reported the situation and arranged for the reserve platoon from FOB Delhi to deploy to assist him. Lance Corporal 'Alpha' Thomas produced an aluminium assault ladder which was used as a stretcher on which the badly wounded Davis was placed. Blaney was placed onto a stretcher but when this proved difficult to manoeuvre in the terrain, Drill Sergeant Darren Chant who had accompanied the patrol hoisted the still bleeding soldier onto his powerful shoulders. Morgan showed incredible courage and determination as, walking on a badly broken and bleeding leg, he gritted his teeth through the pain and plodded on, led by just one man and unable to see through his damaged eyes.

Colour Sergeant Barnett, the mortar fire controller, had mustered a few uninjured men and had coordinated the return of fire while the casualties were being extracted. He now adjusted the fire of the British mortars in order to discourage the Taliban from following up their attack. Roughley and Tiernan carried the body of Probyn between them; there was no question of leaving him behind. It was a terrible but necessary task and there was no place for emotion just yet. The situation was still very precarious and the enemy fire whipped across the open ground as the shattered platoon struggled towards the Fire Support Group who had continued to cover the extraction. The many bomb craters hindered their progress and the darkness made it difficult for the troops to keep their bearings. The huge blast had blown the night vision goggles off of the helmets of many of the troops and the noise of battle and continual streams tracer ammunition caused further confusion.

Captain Gaunt in command of the Fire Support Group, shouted for a cyalume to be cracked in order to guide in Tiernan's withdrawing troops. Hennell duly obliged and rapidly snapped one of the luminous glow sticks, holding it in the air like a beacon. The light source was quickly spotted, but unfortunately it also attracted fire from the now closing enemy and tracer rounds cracked by near Hennell and Guardsman Darren Poyser. After what seemed like an age, 3 Platoon finally reached the Fire Support Group who were now almost out of ammunition. There was time only to make a brief check on the condition of the wounded men before moving north again. Gaunt's little group maintained a baseline and poured fire into a tree line about 200 metres away from which the enemy were clearly trying to advance. The ammunition situation was now becoming critical and Gaunt screamed for his group to fix bayonets. Mercifully the reserve platoon soon arrived to assist with the casualties and CSM Chris Gilham followed up with a quad bike and trailer. The wounded men were placed in to the trailer and Gilham sped off over the difficult terrain. Shortly afterwards, and just in time, the Fire Support Group were able to break contact and to follow up behind the remainder of 3 Platoon.

The medical officer was again ready and waiting to receive the casualties and the emergency response team was inbound from the distant Camp Bastion by Chinook. Together with the company medics and many other willing hands the casualties were triaged. In the light it was plain to see that all of their injuries were even worse than suspected and urgent evacuation was needed. Tourniquets were adjusted and field dressings were applied over the existing blood-soaked dressings. Both Blaney and Davis had terrible traumatic blast injuries and had lost a great deal of blood. Morgan's shattered leg was only now fully discovered and people were amazed at how he had managed to walk out of the contact area. He was still blind which made this feat even more impressive. It was necessary to move all the casualties to a nearby location where the helicopter could land safely and extract them to the waiting surgical team at Bastion. As the Chinook landed, the wounded men were carried to the chopper on stretchers; it had been 75 minutes since they had been injured. The evacuation had been an impressive feat under fire, over difficult terrain in the dark with several desperate casualties.

The remaining members of the patrol only now had time to take in what had happened. No one knew the full story and some were not even aware that Daniel Probyn was dead. 'Probes' was a popular member of the company, a strong personality whose absence was to leave a large void. Some men were in mild shock, others chain-smoked as they discussed the evening's events. It was only now that Guardsman Ashley O'Sullivan reported his injuries. O'Sullivan had shrapnel wounds to his back and to his left arm, but had kept quiet so as not to be a burden to his comrades. He had helped to carry his friend Nick Davis on the ladder and in the dark no one had noticed his injuries; this was an impressive act of selfless dedication. Meanwhile Padre Dunwoody was, by coincidence, visiting 3 Company and did what he could to comfort those who had lost close friends. No one yet knew the fate of the other casualties who were clearly gravely ill when they were extracted. Few felt the loss as keenly as Tiernan, the young officer had seen his men shredded by the explosion. He was incredibly proud of the way everyone had pitched in to help and the casualties themselves had shown immense bravery. For now there was little that could be done other than administration. Blood-soaked clothing was discarded and the men tried their best to get some rest, but few were able to sleep.

By the following morning the news of Probyn's death had got around the whole company and there was a subdued atmosphere in FOB Delhi. It was to take a while for news of the wounded men to reach Garmsir. Both Blaney and Davis had been rushed into emergency surgery at Bastion and it was subsequently necessary for both to have their right legs amputated. This was a further blow to the company, but they were relieved that the casualties had survived the incident. It was a new day and the Taliban were still out there waiting to strike. The company fighting spirit was still present and many vowed to avenge Probyn's death. They would have ample opportunity in the months ahead.

12

SPREADING INFLUENCE

The terrible news of Daniel Probyn's death was broken to those in FOB Tombstone by Lieutenant Colonel Carew Hatherley on the morning of 26 May and it was another dreadful blow. In FOB Robinson, Major Toby Barnes-Taylor relayed the news to the newly arrived men of the Inkerman Company. 'Probes' had been a proud member of the Ribs before joining 3 Company just a couple of months earlier. He had many friends in the Inkerman Company and it was a real shock to hear of his sudden death. The Ribs were told that it might be possible for a few of them to return with 2 Company in order to attend Probyn's repatriation ceremony, but no one took up the offer. Everyone felt that they had an important job to do in Sangin and that was what 'Probes' would have wanted. It was certainly a bad way to start this new mission in Sangin and, like their comrades in Garmsir, the Ribs were determined to pay the enemy back.

Over a period of only 24 hours, Major Marcus Elliot-Square's men took over the Sangin PBs from the tired soldiers of 2 Company who were looking forward to returning to FOB Tombstone and a well-deserved rest. Preparations were made for the journey back to Camp Shorabak and 2 Company and 2nd Kandak departed in convoy shortly after dawn on 27 May. The Grenadiers were more than aware that the Taliban had turned their frustration at losing control of Sangin to placing mines and IEDs along any likely routes. Barnes-Taylor had planned his route carefully but the 39-vehicle convoy was both slow-moving and vulnerable to enemy attack. As usual there were a mix of heavy, lumbering 6 x 6 trucks, ANA First Rangers loaded with enthusiastic soldiers and the ever vigilant 2 Company mentors in their British military vehicles.

After several hours driving in the intense heat and constantly slowing to a stop in order to allow the Afghans to catch up, everyone was fairly tired. The convoy had negotiated many wadis of varying dimensions and there was no sign of enemy activity. Yet another dry stream bed presented itself to the leading British vehicles and they cautiously halted to look for any sign of disturbance that might indicate a concealed mine. After several tense minutes of searching the surrounding area, the waiting vehicles were given the go ahead to cross and the WMIK leading the long train of vehicles rolled carefully down the little incline into the sandy depression and then climbed safely and steadily up the slight rise at the other side. Each of the British vehicles followed , being careful to maintain their distance from the vehicle in front in case anything went wrong. After about ten minutes of slow manoeuvring all of the British vehicles were safely across the wadi and the lead WMIK waited patiently some 200 metres further on so that the Afghans would not be left behind. As the British troops chatted and waited for more vehicles to cross the obstacle there was a sudden huge explosion, the shockwave from which was clearly felt by those in even the lead vehicles. The first Afghan vehicle to cross the wadi had triggered a huge mine and large chunks of tangled metal now rained down, narrowly missing the alarmed passengers in the vehicles that had already crossed. Only two vehicles in front of the explosion, Staff Sergeant Dave Harrison ducked instinctively as the windscreen on his Pinzgauer was smashed by flying stones and Colour Sergeant Walker, who was driving, asked if it was his own vehicle that had been hit. So powerful was the explosion that the crews of several British vehicles were convinced that they had triggered the explosive device themselves. A heavy vehicle bumper landed with a thud directly in front of Walker and Harrison's Pinzgauer. The smell of cordite from the explosion was carried on the wind and provided confirmation of what had happened, if any were needed.

Looking back along the line of vehicles, the Grenadiers could see the twisted wreck of a Ford Ranger in the wadi, smoke still rising from its mangled chassis. The ANA troops were quick to dismount from their vehicles and they ran to the wrecked ranger to assist. The British were more considered in their approach, there was an obvious danger of more mines and no one would be helped by detonating another. Captain James Shaw shouted at the Afghans and did his best to make them stand still but to little avail. Eventually the ANA soldiers started to fan out and to look for secondary devices. Sergeant Brooks moved rapidly on foot to the scene, using the vehicle tracks on the stony slope to reach the stricken vehicle. The front of the ranger and the engine had been almost completely destroyed and the vehicle's cab was crumpled like a toy. Brooks quickly discovered that there were three Afghan casualties, two of them very serious; the driver looked to have badly broken and mangled legs and a severely damaged arm and the passenger had some serious and ugly-looking injuries too. Brooks was joined by CSM Darren Westlake with other Grenadiers and he organised the first aid to the gravely injured men. He assessed that their chances of survival were slim, especially the driver. Captain James Fox coordinated the movement of the MERT from Bastion to collect the casualties, as dressings and tourniquets were applied to their wounds. The driver had lost a great deal of blood and Brooks tried hard to stop the bleeding. While the British provided first aid, the Afghans helped as they could, but there were still many ANA soldiers roaming freely around the wreck and the Grenadiers were nervous that another device could be triggered. The casualties were loaded onto stretchers and mercifully the Chinook arrived within the hour, carrying the desperately-needed medical teams on board.

As the dull thud of rotor blades faded on the wind, Westlake ushered the Afghans back into their vehicles. Nobody was quite sure how seven British vehicles had managed to drive across the wadi without triggering the mine. Large lumps of shrapnel and engine parts still littered the ground around the convoy and no one was more aware of the near miss than Captain Dave Groom, who had not long since narrowly escaped a mine strike with the Americans.

The long snake of overloaded vehicles once again started south. Keen eyes scanned the route ahead for any signs of disturbance and the laboriously slow journey continued. It took more than nine hours to cover the distance from Sangin to Camp Shorabak, and when they arrived everyone was exhausted, apart from the excited Afghans who were eager to depart on leave in the following days. It had been a tough few weeks in Sangin for 2 Company. They had been fortunate not to take any casualties and the mine strike had been a timely reminder of how vulnerable they were to these indiscriminate and deadly weapons.

The following day Hatherley escorted General Muhayadin Gori to the field hospital to visit the casualties that the ANA had taken in the recent action. The ANA troops were treated to the highest standards and received the same care as their British colleagues. Despite the best efforts of the surgical teams, the driver from the mine strike sadly lost his life, but the passenger survived. This was due in no small part to the first aid efforts of Brooks and others.

When British troops were killed in action it was customary for a repatriation ceremony to be held at the airfield in Camp Bastion. Those who gave their lives in the service of their country were carried to the waiting aircraft on the shoulders of their comrades so that they could be returned to their families in a dignified fashion. These ceremonies were sadly becoming more frequent and, by the end of May, 12 Mechanised Brigade had sent the bodies of six men home to their loved ones. There were no such ceremonies for the Afghans and the Grenadiers considered this to be a great oversight. It was suggested to Muhayadin that some form of ceremony be devised and the General enthusiastically agreed. For future ANA fatalities, the OMLT troops would assist the Afghans in arranging small culturally acceptable ceremonies at Shorabak. As far as the British troops were concerned, it was right and proper that the British should recognise the Afghan sacrifices as well as their own. Muhayadin was often seen attending the British repatriations in a public display of solidarity but the truth was that the ANA casualty rates were usually even greater than the British. For the Afghan wounded there would be no compensation or rehabilitation if they recovered from their injuries and an amputee or a soldier with a permanent disability would face a life of terrible hardship. The OMLT troops were becoming aware of just how difficult life was for the Afghan soldiers.

As 2 Company cleaned themselves up and enjoyed their first decent meal in many weeks, the Inkerman Company were experiencing a very different transition in Sangin. The three small teams now dispersed to the PBs busied themselves with the defence of their little outposts just as their predecessors in 2 Company had done. There was no running water and no fresh rations for these men. The British troops cooked over open fires like their Afghan charges, but ate mainly from sealed tins or packets of processed food to reduce the possibility of sickness. Each man had his own camp bed complete with compact mosquito net. Bottled water, along with ammunition, was the most precious resource and this was usually delivered by CQMS Day and company headquarters during one of their frequent expeditions from FOB Rob. The remaining troops from 3rd Kandak together with their mentors were held in FOB Rob as a reserve force that could be used to support operations or to react to enemy attacks on the other static locations. The IED threat, particularly on Route 611, continued to worsen and any excursion out of the FOBs was fraught with danger. The Taliban had also stepped up the rocket and mortar attacks on FOB Rob in order to harass the coalition troops there. The Ribs did their best to fortify the vulnerable perimeters of their new homes and patrolled regularly in order to disrupt enemy operations. The troops in the FOBs often joined up for joint patrols. Captain Phil Hermon and his men in PB Tangier regularly saw Captain Vince Gilmour from Waterloo and Sergeant Summerscales from FOB Blenheim to the north.

Although Operation Silicon had only just ended, another major operation had begun to the south. Operation Lastay Kulang was another task force offensive operation. Operation Silver had brought some form of stability to Sangin, although by no means complete security, and Operation Silicon had done a similar job to the south in the Gereshk area. The villages between the two towns were still heavily occupied by the enemy, as the BRF had found to their cost when they probed these areas. This meant that the route between the towns was perilous indeed. The Task Force Helmand plan was for the settlements in this area to be probed in force and for any enemy activity to be dealt with by a manoeuvre operations group, which would again be based upon B Company of the Royal Anglians. This group was to be self-sufficient and would drive to FOB Robinson, effectively fighting their way up the valley whenever they met the enemy. The operation was expected to take about six days and the group would target a series of suspected enemy strong points along the way. The group would have considerable firepower and would contain some light armoured vehicles and the logistic support necessary to sustain it in the desert for the duration of the operation. The second phase would see them – together with the troops already in Sangin –

attempt to clear the area to the north of the town. After Operation Silicon the Anglians had withdrawn to Camp Bastion in order to prepare for this new operation. The Queen's Company were still deployed in the PBs north of Gereshk and had no warning that they might be involved in the mission.

Lieutenant Paddy Hennessey was unconcerned when he was called into the joint command centre at Gereshk with the rest of his callsign. Hennessey was expecting to be given orders for a low level patrol task in the area of the Gereshk bases and was shocked when Major Martin David briefed him for his future involvement in Operation Lastay Kulang. A late decision had been taken to involve the ANA in the operation and the task had fallen to 1st Kandak. Amber 63 were to move directly to FOB Price where B Company and the manoeuvre operations group were conducting their final preparations for the operation. Hennessey's platoon sergeant, Colour Sergeant Yates, who was attached from the Argylls, was incredulous at the short notice. B Company had been preparing for the operation in Camp Bastion for a week, but Amber 63 had received only a few hours' notice for a complicated and dangerous mission. There was a great deal of unrest and initially some moaning, but Yates and Lance Sergeant Rowe quickly organised the platoon and sorted out the administration and maintenance required. The more experienced Afghan soldiers from the heavy weapons platoon were selected together with their vehicle-mounted DShK machine guns. If there was going to be a fight, Hennessey was going to have the best firepower available to him. Amber 63 made their rendezvous with the manoeuvre operations group in good time and with their 1st Kandak soldiers equipped for six days.

The following three days were bloody affairs. The infantry troops from the manoeuvre operations group were ferried to the outskirts of the target villages in the Viking armoured vehicles of the Royal Marines armoured support troop. Once there, they dismounted and infiltrated on foot. Invariably, enemy resistance was met and there was some vicious close quarter fighting. 1st Kandak and their OMLT mentors remained mounted in their vehicles and gave support wherever they could. The enemy often engaged from the western bank of the Helmand River and it became obvious that they were targeting the British vehicles with indirect fire. Lance Corporal Price had a nasty shock when an enemy mortar round landed a little too close as he was answering a call of nature in a ditch. The Grenadiers looked skywards as fast jets and attack helicopters swept in to support the ground troops. The enemy took a hammering, particularly from the air, but there were casualties in the manoeuvre operations group too and the familiar beat of Chinook rotor blades could be heard overhead as casualties were extracted. The fighting was particularly vicious around Mirmandab and Hennessey's 12.7mm DShKs fired wherever they saw the Taliban.

This pattern was repeated for three days as the manoveure operations group slowly advanced to contact up the valley. The heat and combat intensity took its toll on the soldiers and even the ANA troops seemed exhausted. In one regrettable incident during a night move, an ANA soldier accidentally shot a colleague travelling in the vehicle in front. The wounded man was lucky to survive. The going was tough in the harsh desert terrain; the fine sand repeatedly slowed the British vehicles and the tracked Viking personnel carriers were frequently used to tow the less mobile Pinzgauers. Progress had been painfully slow so there was great relief when Sangin and FOB Robinson came into view; the troops would be able to rest up in the relative security of the FOB. Tragically, one of the Vikings that had done such vital work during the operation struck a mine, seriously injuring the Royal Marine crew in the cab. The wounded were efficiently extracted and the British and Afghan convoy rolled into the FOB to rest and reorganise themselves.

*

The BRF had been active throughout all stages of the operation. They had probed numerous villages and conducted detailed reconnaissance of likely enemy positions, and they frequently became involved in serious firefights. Their deployment had been brought forward based on intelligence that there had been an uprising against the Taliban in the village of Putay. The whole of the BRF were told to make best speed to support the rebellion which sounded like good news for ISAF and government forces. If the local civilians were successful in defying the Taliban it might encourage further anti-Taliban activity. Having driven through Sangin, Major Rob Sergeant's men arrived in an overwatch position. Everything seemed to be quiet and a local civilian was questioned on the situation in the village. He was able to confirm that there had been some resistance to the Taliban and that this was led by the local schoolmaster. The so-called rising had lasted for about 20 minutes. The Taliban had brutally beheaded the courageous teacher and the resistance had evaporated. It was decided that as they were already in place the BRF would enter and clear the village. Most of the night was spent chasing the enemy around the surrounding area, but the Taliban were in no mood for a fight and they disappeared into the countryside. In the following days there were further noisier encounters with the enemy in the villages of Baluzay and Jusyalay. Lance Corporal Rob Pointon was hit by shrapnel during a fierce engagement in the latter. The young NCO was eventually evacuated but was able to return to duty later.

Back in Sangin, the Inkerman Company were settling in and over the following days they deployed in support of Operation Lastay Kulang, and prepared to clear some of the troublesome areas to the north of the town. It was now the turn of Captain Ed Janvrin and a small OMLT team from the Ribs to support the ongoing operation. On the evening of 27 May, Janvrin delivered his orders to the small detachment by torchlight and the troops then set about their last-minute preparations. Early the following morning Janvrin and his troops drove across the dust bowl inside the FOB where they could see the manoeuvre operation group with their powerful Mastiff armoured vehicles forming up in their order of march.

The OMLT troops were tasked to clear the route through the town to the district centre. The three OMLT vehicles halted at the dry wadi crossing point south of the town. This area was notorious for IEDs and a very careful clearance was conducted. The Afghan soldiers may not have fully understood the complexities of this drill but their eyesight was exceptionally keen and the British troops trusted them to spot any disturbed ground. Nothing was found and the Grenadiers moved into the town where the ANA troops mingled with the locals, gathering intelligence. When the route was secure the Afghans moved in and out of the shops in the bazaar and sat with the men drinking chai. This interaction with the population was absolutely vital to build trust and to measure the general feeling in the town. The ANA did this in a way that foreign troops could not possibly hope to achieve. Nevertheless, the Grenadiers were very aware that it was important to win the hearts and minds of the civilian population.

Lance Corporal Tim Leatherland had taken the time to find out what the Afghan children wanted the most and to his surprise he discovered that ballpoint pens were a valued object. The Afghans desperately wanted their children to learn how to read and write, in defiance of Taliban aspirations. Leatherland's girlfriend had been instructed to send out boxes of black ballpoint pens and these were now stowed away in the young NCO's pack. He dished them out to the children who clamoured to get hold of one of these prized possessions. The fact that Leatherland had done this on his own initiative was evidence of the commitment shown by the British troops. As they sat smoking and drinking chai with the uniformed ANA men, the civilians noted that something out of the ordinary was taking place. Most of the local men were not Taliban but it was certain that the news that something was going on would soon reach the enemy. After what seemed like an age, the B Company convoy rumbled up the road which was lined at intervals by Grenadier and ANA troops. As the huge armoured Mastiffs passed by, it occurred to some of the Grenadiers that the ANA and OMLT had just secured the route in light vehicles for the much better protected personnel carriers to pass. The convoy moved safely into the district centre and the Ribs, led by Captain Alex Corbet-Burcher in his WMIK, followed.

The centrepiece of the district centre which was located on the western side of the town was a number of large three- and four-storey flat-roofed buildings. These had been heavily fortified with sandbags and the flat roofs were surmounted by bunkers and firing positions. Camouflage nets blew in the wind and the menacing barrels of GPMGs and .50 cals protruded from the bunkers. The grey walls were pockmarked with bullet and shrapnel holes. Some of these dated back to the previous summer when the men of 16 Air Assault Brigade had been virtually besieged here. Radio masts were staked to the ground outside the buildings and smaller ones were lashed to the roofs to improve communications. On one side of these buildings a large walled garden had been turned into a helicopter landing pad. A tributary of the Helmand River flowed under the perimeter wall where it was forced through a small arch-way. Inside the walls the river was deep and fast-flowing, but the bank next to one internal compound was shallow and it was possi-ble to wade in and bask in the refreshing waters. On that day, several men were waist-deep in the cool water and others appeared to be washing their clothes. Halfway down the flowing river a flimsy aluminium bridge had been positioned for troops to safely cross over, albeit one soldier at a time. Once the Grenadiers had dismounted from their dusty vehicles they were also treated to the sight of Ross Kemp and his crew filming the battered interior of the base. The troops were allowed a few hours respite inside the secure walls of the district centre before moving out once again for the next phase of Operation Lastay Kulang.

The second major phase of the task force plan involved a clearance to the north of Sangin. Since being pushed out of the town the Taliban had consolidated and it was thought that most enemy activity originated in this area. Two companies from the Anglians with support from the manoeuvre operations group were to clear through this dangerous area in order to loosen the enemy's grip around Sangin. The Inkerman Company men together with their Afghan colleagues were to support the operation. Their first task was to secure the designated start line for the infantry advance to contact. Janvrin and his men drove out of the district centre, through the bazaar and onto route 611. As they headed north and passed FOB Blenheim they got a wave from their Grenadier colleagues there. This was the last point held by coalition troops until the road reached Kajaki, a long way to the north.

On the western or left side of the dirt road the ground fell away to the Helmand River. The vegetation here was a lush green; there were poppy fields, crops and thick tree lines, all excellent cover for the enemy. On the other side of the river the rocky ground rose again until the vegetation fell away to be replaced by sand. To the right of the road the ground was comprised of barren gravel interspersed with mud-walled compounds. Standing in the cupola of the rear WMIK, Lance Corporal Eddy Redgate traversed the big .50 cal so that it pointed towards the Green Zone and Lance Corporal 'Pez' Perry in the lead vehicle did likewise. As the little convoy got further away from Sangin the sense of apprehension grew; British troops had rarely ventured this far out of the settlement. The civilians in the town were used to NATO troops and took very little notice, but out here the locals stood and stared, a clear indicator that this was Taliban country. The clearance operation, which was planned to commence early the next morning, was to be centred on the villages of Jusyalay, Putay and Garm Ab. These hamlets were tucked away behind thick vegetation near the river about 7km from Sangin. The ANA were to advance north-west on the left flank of B Company, clearing the south-western part of Jusyalay, once they reached the river they were to swing south and all Taliban in the area were to be cleared from the many farms and compounds.

When the OMLT and ANA troops arrived at their destination, some way short of the villages, they orientated themselves and set about clearing the immediate area. An imaginary line was secured; this would be crossed in the morning by the British infantry troops and it was important to ensure that no Taliban were able to disrupt the advance before it even got started. Janvrin's little group consisted of ten OMLT troops and about a hundred ANA soldiers. The Inkerman Company were thinly spread and hardly knew their Afghan charges at this stage. It was an uncomfortable feeling to know that the nearest NATO troops were 7km to the south and large numbers of Taliban were very close. There was a short period of tension when the ANA reported enemy activity in a nearby compound, but a clearance patrol soon proved this to have been a false alarm. As darkness fell the Grenadiers posted the ANA sentries, but the feeling of isolation didn't leave them. The Ribs placed one of their own light machine guns on a compound roof just off the road and positioned their own sentries. The ten UK soldiers spent a nervous night manning the rooftop position and snatching some sleep closer to the road when they could. Just before first light everyone was roused and they manned stand to positions. The Anglians would be arriving soon and it was important to ensure that the start line for the operation was secure. The Afghan troops were now also expected to prepare to advance alongside the British.

The Royal Anglian Group arrived at the line of departure a short while later. They had spent the night at a desert leaguer and had dismounted from their armoured vehicles shortly before reaching the road. The troops now spaced out and headed towards the Green Zone, the lead platoon in an arrowhead formation. This was a classic British advance to contact, something that is rehearsed endlessly in infantry training. From a roadside trench Leatherland looked on and thought what an impressive sight a whole company of heavily laden infantry troops made as they walked slowly towards the enemy concealed in the vegetation to their front. Unfortunately the ANA troops were not ready to advance. This was catastrophic news for Janvrin who was furious. The Afghans went about their administration and further badgering was pointless.

Janvrin had the embarrassing job of relaying the news to the Royal Anglians who were already moving towards the enemy. They didn't move far before encountering the opposition; as the advancing troops broke a treeline about 800 metres away, the firing began. At first a sudden burst from the enemy and then the louder and heavier weight of fire from the Anglians as they responded. The heavy Mastiffs and Vikings now moved up onto the road near the OMLT in order to provide fire support with their heavy weapons. The ANA troops – now ready – were chattering excitedly and pointing their weapons towards the enemy positions. The Grenadiers ran up and down the line of Afghans passing the message for them not to shoot.

The Anglians had cleared through the thick vegetation and the partially concealed compounds and had pressed on towards the river. They now swung north once again to force the enemy out. Janvrin's ANA troops were to move down through the Anglians and were to fight through to the river, then swing south along the river bank in the direction of Sangin. The OMLT troops were surprised to hear that they were to do this in their vehicles. They would be able to move quicker but their mobility in the Green Zone would be much reduced and they would be vulnerable to close quarter attack. Now motivated, the ANA troops sped off into the Green Zone. Janvrin, with his Afghan officers, went to locate the B Company commander in order to ensure that each was aware of the other's intended actions.

Janvrin now led the ANA into the Green Zone to commence their clearance. He left the vehicles on the track to provide fire support but most of the work would be done on foot. Before long a few RPG rounds were fired in the direction of the vehicles, exploding loudly. Janvrin and Corbet-Burcher dismounted and moved rapidly through the vegetation with the Afghans. In contrast to their early refusal to advance on time, they now moved fearlessly and efficiently through the compounds and fields of crops. The Taliban fought a determined rear guard, shooting and withdrawing to safety before the ANA closed the trap on them. It was as though they recognised that the fast-moving and aggressive Afghan soldiers would be quicker to close with them. Many of the engagements were at close quarters – under 50 metres – and a fair number of the enemy were killed by the ANA. The rest of the day was spent pursuing the enemy through the maze of irrigation ditches, crops and mud-walled compounds. By the end of the day the ANA had cleared 2.5km of this difficult terrain, more than three times that cleared by their British counterparts. Janvrin felt that the ANA had salvaged their credibility after a disappointing start to the operation. He was highly impressed with their aggressive and rapid momentum in the assault. The OMLT troops were often heard to say that their Afghan colleagues were like children; sulky, stubborn and unmovable at times, but enthusiastic, fearless and effective at others. The trick was to motivate them correctly in order to ensure the most effective response. The Mastiff squadron secured a desert leaguer for the night and the troops withdrew into this secure area for some rest. The OMLT men settled down in an old graveyard and pondered what would happen the next day.

The following morning saw the troops once again clearing the ground between Route 611 and the river. The Taliban had for the most part decided not to stand and fight, but there were still some serious engagements. The OMLT vehicles crawled along the tracks observing keenly for the enemy. When the WMIKs came under attack, Perry and Redgate poured fire into the suspected enemy firing point as Sergeant Tosh De-Vall pumped 40mm grenades from his under-slung launcher at the same point. The enemy fire ceased and a rapid follow up showed that the Taliban had decided to withdraw. This was fairly typical of the engagements that day, but Janvrin and his men pressed on through the highly dangerous thick vegetation. About half an hour later Janvrin reported that the enemy were to the front in the area of the river and he suspected that they were trying to escape to the relative safety of the far bank. He quickly collected a small detachment of Grenadiers together with De-Vall and his Javelin equipment. The little group moved rapidly to within sight of the river where they were able to see a small boat full of escaping enemy fighters, which was just reaching the far bank. De-Vall set up his missile launcher and having received permission to engage he fired the deadly missile. There was a clunk as the firing button was depressed and then a loud whoosh as the missile's motor sparked into life. After a pause there was a terrific explosion on the far bank as the enemy fighters disappeared amid the black smoke surrounding the point of detonation. There were hoots of delight and congratulations for De-Vall but Janvrin had identified more enemy positions to the west side of the river and he now called in mortars and artillery to neutralise the threat that they represented. The ANA troops watched the British artillery wreaking havoc with the Taliban on the far side; they were clearly impressed by the awesome display of firepower.

While this was taking place on the river bank, the other half of the OMLT detachment was still with the vehicles in an open field. Leatherland was approached by the interpreter who quickly passed on the information that the Taliban had identified the ANA vehicles in the field and were apparently preparing an attack on them.

Leatherland launched into action and dashed off to a nearby section of Afghan troops who were resting in the shade of a tree line. The young NCO had the interpreter relay the worrying news to the Afghan soldiers who showed no interest whatsoever. Leatherland asked the interpreter to ensure that the ten or so soldiers understood that the enemy might be preparing to attack. The Afghan commander simply lifted his sunglasses and replied through the interpreter, 'Yes, we know. We are waiting for them.' Leatherland was staggered by the lack of interest; some of the men looked as though they were about to go to sleep. After much badgering, the ANA section reluctantly raised themselves up and took up fire positions.

Back on the river bank Janvrin had now called in air support. It was clear that scores of enemy fighters had escaped to the far bank and were attempting to take cover from the artillery. Aircraft would be able to spot the fighters and direct their bombs more accurately than British guns. The young officer gave his target indication to the aircraft but the pilot did not release his bombs onto the location he had been given. The flyer had spotted enemy fighters hiding in a series of caves a little further away than Janvrin's original indication. The aircraft launched its bombs onto the cave entrance to devastating effect and the pilot reported that he had neutralised the target. On the ground everyone was delighted. By talking to the pilot, Janvrin was able to estimate that about a dozen Taliban had been killed. The following day, the Grenadiers learned from the locals that the caves had also contained a number of civilians who had perished alongside the enemy. The Taliban had as usual used them for cover. It was terrible news which appalled Janvrin and the other British troops; they had always taken the utmost care not to endanger civilians. This was one of many dark moments for the troops and inevitably detracted from an operation which had otherwise been a success and where they had worked well with the ANA.

Over the following days the operation started to wind down. The British and ANA troops were consolidating in the areas they had cleared and the Taliban had withdrawn to lick their wounds after some severe losses. Colonel Rassoul's job was now to try and win over the local elders and to build confidence by showing an Afghan face in the areas that had until recently been under the brutal control of the Taliban. Janvrin took over temporary command of the Inkerman Company as Elliot-Square returned to the UK on leave. Over this period a reappraisal of the OMLT/ ANA deployment was carried out. It was decided that the troops needed to be more in touch with the people and FOB Rob was becoming increasingly distant from what was now the new front line in the area of Jusyalay. By moving the Inkerman Company HQ into Sangin district centre the company would be much closer together and more ANA troops would be available for undertakings in and around the town. The district centre was on the west side of Sangin but was relatively central to the other PBs.

Rassoul set up his HQ in the building adjacent to the OMLT and from here he planned a series of shuras or meetings with the local elders of Sangin and the surrounding hamlets. It was important to hear from them what their priorities were for assistance. Medical support, schools and irrigation ditches were all usually on the agenda. The charismatic Rassoul usually sat cross-legged on the floor surrounded by bearded old men in turbans, who often complained to him about the lack of support they received. It was clear that they had no particular allegiance to the Taliban, but they were frightened to cooperate in case the brutal enemy fighters returned. Security was the key to building confidence, but every patrol that left the comparative safety of Sangin or one of the bases was in grave danger of attack from the Taliban who crept back into the area by night to place their deadly IEDs. Rassoul was often seen strutting around the compound talking rapidly into his mobile phone. It was clear that he had contacts everywhere and that the ANA were vital to success in the area. Most of the fighting had now stopped, but the operation was continuing in its final phase which was to secure this area and to win the cooperation of the civilian population by building their confidence.

During this early part of June the enemy were trying very hard to dislodge the British and Afghan government troops from their isolated PBs. Sangin district centre was rarely attacked but the smaller and more isolated PBs inhabited by the Grenadiers were frequent targets. Small groups of men constantly fought from the rooftops and walls of the tiny compounds to repel the Taliban attacks. Most of these attacks were designed to harass and cause casualties rather than to overrun the little outposts. The enemy fighters crept to within range, fired their RPGs and machine guns and maintained the fight until they thought the British mortars or artillery would find them, they then melted back into the safety of the Green Zone. Coalition troops could not be everywhere and the enemy often moved quite freely, especially at night when they buried their lethal explosives at the roadside. For the OMLT detachments, which sometimes consisted of no more than four men, these were testing times and nerves became frayed.

Rassoul knew too well that his troops were thinly spread in the area and it was difficult to provide security for all of the villages to keep them free of Taliban intimidation. The answer was to convince the elders to raise local militias that could be armed and could operate in the local area supported by the ANA and the British. Rassoul befriended one of the local elders who seemed like a good man with the interests of his community at heart. Some progress was being made and the Afghan Army was doing its best to improve the quality of life of the local people. Unfortunately, during a visit to relatives the cooperative elder was abducted and murdered. The message from the Taliban was very clear. Faced with such brutal intimidation of the civilian population, winning the locals over was going to be a monumental task, but the ANA and OMLT troops continued their efforts. A very useful asset was the light wheeled tractor, simply referred to by the troops as a 'digger'. This little vehicle was operated by a Royal Engineer driver and was capable of digging irrigation ditches for the locals. Such small tasks were hugely significant to the farmers who would otherwise spend weeks digging with picks and shovels. The problem was that the 'digger' was very slow moving which made for some very nervous journeys. To make matters worse the OMLT and ANA were now operating in areas that until recently had been 'Taliban Central'. One of these missions saw elements of the Inkerman Company escort Rassoul and his men to an area just south of Kajaki Sofla. NATO soldiers had rarely been seen here, but US 82nd Airborne troops had recently carried out a huge helicopter assault in the area. Some ANA troops had also taken part, mentored by Lieutenant James Harrison and four men from the Inkerman Company. As expected there had been some fierce opposition and a US Chinook was lost during the operation with the loss of five American and one British soldier. The US troops had now returned to their base at Kandahar and the Taliban had come back to exact revenge and dish out punishment to any civilians that they judged to have collaborated with the NATO troops. Janvrin and his men saw at first hand the results of Taliban retribution. They encountered an old man who had been beaten black and blue; he told them of terrible beatings and beheadings in the villages to the north of their current position.

Rassoul had promised the local elders that he could arrange for irrigation ditches to be dug for them, and the digger, having made the perilous journey north, set about the task. The Grenadiers and the ANA troops fanned out in all-round defence to protect the digger and its Royal Engineer operator. Colonel Rassoul, who was now content that he had fulfilled his promise to the elders, decided to take a swim in the cool waters of the Helmand River. The

Grenadiers watched as the Afghan washed in the swirling waters and even produced a bottle of shampoo which he rubbed into his dark hair. Beyond Rassoul the Afghan sentries eyed the ground to the north and it wasn't long before movement was seen in a distant tree line. A short time later the ANA and Grenadier cordon was in contact with the enemy once again. The ANA brought up one of their rangers with its pintle-mounted DShK and started to fire at the distant enemy troops who were replying in kind. The Grenadiers were now treated to the bizarre sight of Rassoul performing a backstroke in the river as if nothing was happening. The big DShK boomed away, spitting tracer into the distant trees, but the Afghan commander showed absolutely no concern for the battle raging around him.

In the meantime the Inkerman Company men moved to support the ANA. De-Vall launched several Javelin missiles towards the Taliban positions and the GPMG and .50 cals were used to great effect. The enemy eventually melted away and the shooting finally stopped. The Royal Engineer completed his task and the now clean and refreshed Rassoul announced that everyone was very happy with the newly dug irrigation ditches. The return journey was very nervous and everyone expected that the enemy would be waiting for them. The situation was compounded by the painfully slow pace of the digger. Rassoul became increasingly frustrated at the very slow speed and after a short period the ANA vehicles sped off into the distance leaving the Grenadiers and the digger in a cloud of dust. Nerves were tested but the little convoy arrived at their destination without further incident.

It was necessary for the OMLT troops operating to the north of Sangin to secure a base from where they could operate. The Ribs had occupied another compound about 7km north of the town; it was close to Route 611 and had some decent views to enable the troops to see any approaching enemy. The BRF had recently used this location as a base from which to operate. Everyone was aware that they were extremely isolated and that they lacked the combat power to peg back the Taliban in the long term. The surge operations mounted by the task force were highly effective but once the large numbers of British troops were withdrawn from the area, the OMLT and ANA were left somewhat exposed. Support was available in the form of artillery and air cover but these were reactive measures. Although their task was incredibly dangerous the Grenadiers understood the importance of the OMLT role. The future of Afghanistan was in the hands of the Afghans and the ANA would one day take over from the NATO troops currently providing the government with some breathing space. For now they would have to get on with the job at hand.

The BRF had also been ordered north and were now operating in the dangerous area where the Helmand River met with the Musa Qala and Kajaki valleys. The terrain here was rocky and rose sharply into steep-sided hills overlooking the river. A number of observation posts were mounted high on the hillsides. These were ideal locations from which to view the area using their powerful observation devices. The vehicles were concealed in a muster in the low ground and the whole area was protected by sentries and Claymore mines. After two days in position a local goat herder passed by the muster and was stopped by the sentries. The man was recognised as a local who had been helpful to the BRF before and he was questioned about enemy dispositions and strengths. The man proved to be cooperative and a mine of information. He took the BRF soldiers onto some high ground overlooking a local village and from here he pointed out all of the Taliban positions including sentries and the HQ which was apparently located in the mosque. The goat herder was sent on his way and a further day was spent observing the enemy routines. The Afghan's information proved to be entirely accurate. It was possible to identify sentry positions, the enemy strength and routine.

At 0315 hours on the following day the Fire Support Group positioned themselves in a nearby graveyard overlooking the Taliban positions. The assault groups too were in position and ready to go when the order was given. At H-hour the five enemy sentries were killed by snipers and all hell broke loose. Heavy machine gun fire rained down on the enemy positions and the assault troops moved rapidly through the village. More Taliban were killed as they fled the village, others seen escaping along the river bank were soon stopped by 81mm mortars. Things almost went wrong during the withdrawal when two isolated Taliban fighters belatedly decided to put up a fight. They engaged the BRF troops as they passed, but were quickly silenced by a rapid assault. The fighters who had occupied the mosque had narrowly escaped as their blankets were still warm. It had been a textbook operation and had cost the enemy dearly.

The following day CSM Ian Farrell took out a clearance patrol. He encountered some civilians from the village including the goat herder who had provided the essential intelligence days earlier. Farrell asked how the village was and was told that the Taliban were still present but that they now occupied the little hamlet at first light and left before dark. The villagers were still unable to reclaim their homes.

Early the next day the BRF laid an ambush and watched as 12 Taliban fighters crossed the river by boat and patrolled into the village. It was tempting to hit them immediately but the reconnaissance troops held their fire until the Taliban had settled into a routine. Once they were all in the killing zone the ambush was sprung. Javelin missiles, mortars, heavy machine guns and small arms tore into the occupiers. A seemingly endless stream of death poured into the village until there was no movement to be seen. A clearance patrol moved through the collection of battered compounds and confirmed that there were no more enemy fighters left alive. The villagers reoccupied their homes two days later.

It had been a very successful operation. The BRF now moved on to the next task which was to clear the routes back into Camp Bastion for the deployed elements on Operation Lastay Kulang.

On 9 June Janvrin led another joint OMLT/ANA patrol to the hazardous area north of Sangin. There were three British vehicles: CSM Wayne Scully commanded the lead WMIK, Janvrin rode in a Snatch Land Rover in the centre and Corbet-Burcher brought up the rear in a second WMIK. The ANA followed on in seven overloaded rangers. Janvrin and Rassoul had arranged to meet the local elders in Jusyalay and to hold a shura – an essential part of the hearts and minds battle. Sitting face to face with the locals and talking about their concerns was vital for confidence building. While the command group talked to the Afghans, Scully and the remainder of the OMLT detachment positioned their vehicles in all round defence. Scully scanned the area with binoculars and discussed the likely enemy approach routes with Day who was driving the vehicle. Redgate manned the .50 cal on the second vehicle driven by De-Vall whose Javelin equipment was stowed in the WMIK. Guardsman Tony Downes made a brew as the others observed for signs of the enemy. Just as the brew was nearing completion a powerful explosion echoed around the buildings. The explosion was not in the vicinity of the Inkerman Company men or their ANA collegues, but was close enough to be heard, Scully estimated the distance as being 2km to the south. This placed the seat of the explosion somewhere on Route 611 between PB Blenheim on the northern edge of Sangin and the newly established northern base. To their relief the patrol was able to establish fairly quickly that the explosion had not come from the PB. As the British troops speculated about the cause of the detonation and studied their maps to try and identify a likely location, information started to come in via the ANA radios. It soon became clear that an ANA vehicle carrying water had been hit by an IED. There was no news on casualties but they would obviously need help. Before long Rassoul arrived; he was keen to wrap up the shura and move to help with the stricken water truck. Downes was forced to throw away the freshly prepared brew and to once again man the GMG on the rear of his CSM's WMIK.

The vehicles manoeuvred back onto Route 611 and headed towards the suspected site of the explosion. There were four ANA vehicles now in the lead, closely followed by the small OMLT group. As the convoy moved towards a hamlet the leading ANA vehicles started to slow and then stop. Colonel Rassoul could be seen walking from his vehicle towards a group of locals in a small hut. The British vehicles came to a halt and Janvrin's Snatch Land Rover overtook the lead WMIK in order to catch up with Rassoul and his men. The remaining British troops prepared to dismount and conduct the all too familiar anti-IED drills. Suddenly a huge explosion shook the ground throwing up masses of dust. The rear WMIK commanded by Scully was thrown into the air, flipping upside down in the process. His arms flailed as he found himself fighting through the dust cloud. He landed heavily and was temporarily knocked unconscious. Day, who had been driving, was still in the vehicle as it landed upside down on the gravel track. The vehicle's roll bar kept him from being crushed, but he was nevertheless trapped underneath the now burning vehicle.

High velocity rounds were striking the gravel track and they seemed to be coming from all directions. Redgate, who had been knocked backwards by the force of the explosion now jumped down from his position in the front WMIK. The scene presenting itself was terrifying. Flames were starting to lick out from the upturned vehicle. Redgate had no idea where the occupants were and he feared that they were trapped beneath the burning wreck. Then he saw Scully who had regained consciousness. Scully was clearly disoriented and, as he stumbled around, Redgate ran to him and quickly dragged him behind the cover of his vehicle.

Bullets ricocheted off the gravel nearby. Corbet-Burcher was trying to make sense of what had happened and was shouting instructions to the shocked OMLT men. Turning back to the wrecked WMIK which was now billowing black smoke, Redgate was amazed to see an arm suddenly appear from underneath the vehicle, followed by a head and the wriggling form of Day as he desperately tried to free himself from under the burning wreck. Redgate sprinted across and helped to pull the shaken NCO from the upturned vehicle. Other members of the patrol now arrived at the scene and searched desperately for the still missing Downes, screaming his name and squatting down to look for signs of life from the crackling WMIK.

The Taliban ambush came from various directions and was well planned. Stuck in the killing zone the Inkerman Company men were now extremely vulnerable to the heavy rate of enemy fire that poured in. Rassoul's ANA soldiers, further away from the blast, reacted swiftly and aggressively. They quickly identified the enemy positions and launched an immediate assault, covering the ground rapidly, forcing the enemy to break contact and to start a withdrawal. The rate of fire reduced and eventually stopped as they ran for the cover of the Green Zone. As the ANA secured the area, Corbet-Burcher located Downes. Day assisted and tried to give first aid but Downes was dead. The young soldier had been thrown clear of the vehicle and killed instantly.

Ammunition from the burning WMIK was starting to explode, adding to the hazards faced by the still shocked troops. Janvrin, whose vehicle had only escaped the fate of the WMIK by a last minute decision to overtake, now turned his attention to the fleeing Taliban. He called Hermon at PB Tangier and told him to move up to the contact site with a strong force of ANA. Hermon immediately crashed his men out and headed north on Route 611. At the scene of the explosion Rassoul's men were quick to round up any civilians that they suspected of involvement in the ambush.

It was obvious that the device could not have been dug into the road without the locals knowing about it. Two men were arrested and the Grenadiers were unsympathetic to their pleas of innocence; the Afghans would deal with these men using their own judicial process. Janvrin together with Rassoul decided to mount a follow-up strike into the Green Zone. This was based on instinct and on intelligence gathered by Rassoul's men. It seemed highly likely that their attackers were hiding out close by.

Janvrin, supported by the recently arrived Hermon and his team, took off in pursuit of the enemy. It was a very nervy affair; the Taliban were almost certainly hiding nearby and the likelihood of another ambush was high. There was however no sign of the enemy who on this occasion decided to avoid contact. The chase was brought to an abrupt end when Perry, who was driving Hermon's WMIK, misjudged the width of a narrow Afghan bridge and the vehicle toppled off into the stream below. Perry, Hermon and Lance Corporal Kiddell were all thrown into the cold water, but luckily no one was seriously hurt. The Taliban escaped across the river and after some difficulty the stranded vehicle was recovered and the group returned to the scene of the explosion where the WMIK was still burning.

The survivors from the explosion were only now getting over the initial shock and the realisation that Tony Downes was gone. When an escort of Vector vehicles from Sangin arrived, the battered patrol carefully lifted up the body of their comrade and drove away from the settlement. A helicopter had been requested and this was scheduled to fly into Sangin district centre. The journey back to Sangin was a difficult one but once they arrived the doctor formally confirmed that Downes was dead and provided treatment for the minor wounds suffered by the survivors. Scully, Day, Redgate and De-Vall escorted their fallen comrade back to the hospital at Camp Bastion. Hatherley and Captain Vince Gaunt met the shaken party and the little group sat down and discussed the horrors of the day over tea and biscuits; a curiously calming ritual. The following days were both sad and challenging for the Inkerman Company. Scully remained at Bastion and organised the bearer party for the repatriation ceremony. He then accompanied his fallen comrade on the flight back to the UK where he met the young soldier's parents at RAF Lyneham. It was an intensely emotional experience. Guardsman Downes was a popular and capable man whose absence was now sorely felt in the dispersed Inkerman Company.

For the rest of the Inkerman Company who had remained in the Sangin area there was to be no respite. On 11 June Janvrin was leading another patrol in the area of the Charkakhan wadi when his vehicle struck a victim-operated device, triggering a large explosion and another Taliban ambush. One of the ANA soldiers was seriously wounded and was saved only by the prompt actions of Janvrin's interpreter. It was necessary for the patrol to once again fight its way out of the contact area, which it did effectively. The Taliban were now fighting back hard to harass the British and Afghan government troops at every opportunity. On 15 June Janvrin led another patrol from Sangin intending to visit the various bases to conduct a resupply and to check on the hard-pressed men in the outposts. De-Vall and Redgate had just rejoined the company and the patrol departed leaving only Summerscales, Captain Richard Dorney, and an RMP lance corporal in the OMLT compound. About five minutes after the patrol departed, a massive explosion shook the ground and echoed around the FOB. It clearly came from the direction of Route 611 and the remaining Grenadiers feared the worst. Their apprehension increased when no one was able to raise Janvrin's callsign on the radio. There was frenzied activity as the troops in the district centre desperately tried to discover what had happened. The mystery was solved about ten minutes later when a series of battered ANA rangers sped through the gates, their horns blasting as they skidded to a halt outside the OMLT house. An IED had been detonated against an Afghan patrol inside a police check-point. The police had abandoned the post overnight and the Taliban had concealed a huge device inside the chicane formed by the bastion walls. In the confined space the effect of the IED was devastating. Dorney, Summerscales and the RMP helped to lay the ANA casualties on camp beds where their injuries could be assessed. It quickly became apparent that three of them were beyond help and one of the others had some very serious chest injuries which would require urgent evacuation. There were two other injured men, one of whom was a stretcher case with serious shrapnel injuries. Summerscales and Dorney were joined by some Royal Anglian troops who helped to apply dressings and to carry the wounded survivors to the aid post where they could be assessed by the medical officer.

The Afghan dead were placed in body bags and Dorney confirmed their identity with the ANA company commander. It was important to ensure that the dead soldiers were correctly iden-tified so as to reduce the confusion when they were returned to Shorabak. Islamic tradition dictates that the men should be buried quickly and the Grenadiers took the utmost care to respect the Afghans' religion. An hour later the three survivors, accompanied by Dorney and Summerscales, flew back to Camp Bastion on the MERT Chinook. The Afghan soldier with the chest injuries was very badly wounded and the doctors and nurses aboard the Chinook fought to save his life. At Bastion the helicopter was met by several ambulances which collected the two stretcher cases and an older NCO who was able to walk. They were taken the short distance to the tented field hospital where the incredibly profes-sional British medical staff took over. The three casualties were immediately taken away for assessment while the two Grenadiers were held at the reception area. Dorney and Summerscales unloaded and surrendered their weapons before being provided with a much-needed brew. After about half an hour they were allowed to visit two of the casualties who had been stabilised in the assessment ward. It was gratifying to get a 'thumbs up' from the wounded and heavily bandaged Afghan soldiers. The third man had already been taken to the operating theatre where he was receiving life-saving surgery. Some of the hospital's very dedicated staff were reservists who had volunteered to temporarily leave their jobs in the National Health Service and undertake a tour in Afghanistan. It was reassuring to see how effectively casualties were dealt with.

Back in Sangin the relentless insurgency continued with daily attacks against the coalition troops. The loss of Guardsman Downes had hit the Ribs hard but they had vowed to pay the enemy back and they fought hard every time they encountered the Taliban. The northern base was furthest from the Sangin district centre and most isolated and frequently attacked location. It was now known as FOB Inkerman in honour of its founders. Hermon and his team took over the base just after Downes was killed. The PB was split into two areas: one occupied by the ANA troops and one by the OMLT. A small single-storey building would be home for the foreseeable future and Hermon's little team set about filling sandbags to provide an area from which they could fight from the roof. From here they had a good view of the ANA area but more importantly they had a good field of fire into the Green Zone. The area to the front was fairly secure as the ANA positions guarded this well. The rear of the compound was a different story; this was vulnerable and the OMLT men placed claymore mines and trip flares for additional protection. This proved to be a prudent move as on only their second night in occupation one of the trip flares was set off. The enemy had clearly tried to get in close and at least one man was spotted fleeing on a motorbike. This caused Hermon to reassess the audacity of the Taliban. As a result a double-stacked razor wire fence was erected as additional protection.

During this period of 'bedding in', the OMLT and ANA troops concentrated on securing their new home. As a result, few patrols were conducted and it soon became apparent that in this area the enemy was able to move with impunity. The Taliban were clearly not prepared to allow the British to establish another PB in the middle of what until very recently had been their heartland. In this first week Hermon recorded no fewer than 17 determined attacks on the base. The daily routine of improving the defences and mentoring the ANA continued, punctuated only by the daily and sometimes twice daily enemy attacks. There was precious little down time for the troops but when this presented itself it was used to catch up on sleep and to write the odd letter home. On one afternoon Corporal Justin Huggett, a rugged Australian who was on attachment to the Grenadiers from the Royal Australian Regiment, and Perry were relaxing inside the small building that was now their temporary home.

The afternoon quiet was suddenly shattered by the sound of exploding RPG rounds and small arms fire. The two quickly donned their helmets and body armour; Huggett was fastening his boots when a huge bang shook the room which instantly filled with smoke and dust. Both men's ears rang as they stumbled from the smoking room with their weapons. It had clearly been a close shave but there was no time yet to examine just how close. For the next 90 minutes Perry and Huggett fought alongside their colleagues to repel what was probably the most determined assault on Inkerman yet. The Taliban eventually withdrew and the Grenadiers were able to inspect the little room that Huggett and Perry had vacated. The smoke was now clear and at the far end of the room light could be seen pouring in through a large hole that had not been there earlier in the day. The floor was covered in rubble and their camp beds were coated in dust. Pieces of lethal shrapnel were embedded in the walls and Tosh De-Vall's bed space had been completely destroyed; he would surely have been killed if he had been in the room. Tracing the trajectory of the projectile as it had passed through the walls, Huggett noted that it had passed no further than six inches from Perry's head. The Australian's own mosquito net, sleeping bag and a stretcher stored nearby were full of shrapnel holes. It was a astonishing that neither man had received so much as a scratch. This was all the more so when the OMLT men discovered that it had not been an RPG that had penetrated the walls but a round from a Soviet SPG 9 73mm recoilless anti-tank rifle. These weapons were designed to defeat between 300 and 400mm of armour plate and could reach a velocity of 700 metres per second. Under normal conditions such an event would be enough to cause anyone to think twice about their career choice, but Huggett cheerfully recorded the whole scene on his video camera, together with a lively Aussie commentary. These were dangerous times and the only way to cope with them was with a sense of humour.

The flat roof of the accommodation building with its newly constructed defensive positions was reached by means of a makeshift ladder. Due its height it was necessary to climb the second half of the ladder above the compound walls, which left troops completely exposed to the enemy. There was, unfortunately, no way around the problem and the climb was not for the faint-hearted. De-Vall thoughtfully wrote 'Targets up!' on the adjacent wall as a humorous reminder of the perils facing anyone visiting the rooftop positions.

Fortunately for the detached little group, they were well equipped for a fight. Janvrin had ensured that Hermon and his men had the weapon systems that they needed to defend their base. A tripod-mounted GPMG on the roof ensured that targets could be accurately engaged at long distance. The Javelin missile system could also wreak destruction on any potential attacker and the troops were becoming more and more adept in its use. Sergeant Dan Moore had joined the Ribs too and his expertise with the 81mm mortar now set up in the compound would bring further welcome firepower to the eight-man team. The lethal expertise of the company snipers was put to good use. Any enemy fighter who showed himself long enough to be framed in their sights would not live to attack Inkerman again. Artillery and coalition aircraft were usually available to support the little garrison and Hermon felt relatively secure inside the base even though they were under frequent attack.

Hermon now turned his attention the ground further out from the PB. It was necessary to conduct frequent patrols to gather intelligence and to deter Taliban attacks. The best way to understand the nature of the ground was to see it up close but it was not practical to venture too deep into the Green Zone as the OMLT group lacked the combat power to fight a close quarter battle with a prepared enemy on their own ground.

In late June, Hermon led a patrol out of Inkerman to the north west towards the Green Zone; this was comprised of an ANA platoon and most of his small team. The young officer and his troops moved on foot with a couple of WMIKs following up to provide some fire support and the means to extract any casualties should this be necessary. After only a kilometre the lead Afghan troops came under heavy and sustained fire. Hermon struggled to gain a picture of the situation forward of his position. The ANA troops seemed to be pretty well pinned down and Hermon now tried to manoeuvre to a flank from where he could better support the lead troops with fire. The ground was extremely difficult and supporting fire was restricted by buildings which obscured the enemy positions. The Taliban manoeuvred quickly to counter the threat posed by the OMLT men. On several occasions the concealed enemy tried to flank the ANA troops, a tactic that was well understood and frequently used. After about 20 minutes the Afghan soldiers managed to shoot their way out of their precarious situation and they fought their way back to a base line that the Grenadiers had formed to support them. When the Afghan commander and Hermon were reunited, the ANA officer was furious. It was clear that the Afghans had been rattled by their close shave and their commander now demanded to know why Hermon had not called in an air strike to extract them. Hermon explained that the enemy positions had not been clearly identified and that they were too close to the ANA, apart from which the engagement had so far lasted only 20 minutes. It was unrealistic to expect air cover in these circumstances. The Afghan was still furious and he ordered his troops to withdraw by the shortest possible route. The small group of OMLT men realised that they too would have to withdraw and they managed to extract themselves with artillery support from FOB Robinson. Some aggressive driving and fire support from the standing gunners in the WMIKs helped to deter the enemy from following up. Both the ANA and Hermon's OMLT men returned to Inkerman without any casualties.

Later Phil Hermon confronted the Afghan company commander who apologised for withdrawing so abruptly. It was an apology the British were able to accept as no one had been injured, but some serious lessons had been learned. Liaison and understanding between the British and Afghan troops would have to improve. The Taliban were proving to be cunning and skilful adversaries. They used every weapons system available to attack the PB, small arms, RPGs, anti-tank rockets and even mortars. The enemy was imaginative, very mobile and difficult to tie down. They understood entirely that if the British were able to fix them in a position, overwhelming firepower could be brought down on them very quickly.

The daily attacks on Inkerman continued to mount; these were now referred to by the US Army acronym 'TIC', standing for 'Troops in Contact'. It was hard to think of a more understated term to describe the pitched battles being fought from the rooftop of the PB. The tripod-mounted 7.62mm GPMG as always proved to be an invaluable weapon. It spat bright tracer deep into the Green Zone from its sandbag emplacement and acted as an effective deterrent to movement. The Javelin too sought out the concealed fighters before launching its lethal rocket down onto them. The Javelin didn't need the pinpoint accuracy of the sniper rifles; when the rocket struck it obliterated anything nearby. The black plumes of smoke rising from the Green Zone were testament to its effectiveness.

The Taliban didn't always attack at a time convenient for the defenders and the British often found themselves fighting in shorts and sandals with hastily donned body armour draped over their bodies. On one occasion a naked Moore was seen sprinting from the improvised shower to his firing position. The defenders of the Inkerman compound became very skilled in the use of the weapons in a short space of time. They were also proficient in the use of helicopter support and the ever appreciated strikes by coalition jets, all of which cost the Taliban attackers dear. One disadvantage was that the team got through a staggering amount of ammunition in a short space of time. The intense heat meant that water was also consumed very quickly. Inkerman was frequently resupplied by other OMLT teams from the Ribs and Gilmour and his team often had the unenviable task of collecting water and ammunition from FOB Rob and driving it via the dangerous Route 611 to PB Inkerman. Driving past the site where Downes had perished always served as a stark reminder of the dangers that the patrols faced on this route. The threat from IEDs was now at its highest since the Brigade had arrived and explosions struck coalition vehicles almost daily. When the resupply run troops arrived at Inkerman they were usually greeted as long lost brothers. This was due in part to the fact that they delivered the much needed supplies but there was also a great deal of relief that the patrol had made it to Inkerman without being hit. On rare occasions, members of Hermon's team were able to get back to Sangin district centre and its relative security. The cool waters that rushed through the base provided a special treat and the dusty visitors to the DC were usually found wallowing in the ice-cold water. It was strange how such small luxuries made the dangerous routine bearable.

While the Inkerman men were fighting their daily battles with the Taliban, plans were being put in place to deal a fatal blow to the enemy fighters operating in the area north of Sangin. Operation Ghartse Ghar would once again involve two companies from the Royal Anglians operating alongside the ANA men from PB Inkerman, with their mentors from the Grenadier Guards. The plan involved the various sub-units manoeuvring themselves into a position where they could push the Taliban back into the area to the north-west of Inkerman. If the enemy could be trapped with their backs to the river in a discernible mass, artillery and air power would be used to destroy them. They would also be vulnerable if they tried to escape across the river. Pushing the Taliban fighters back through the Green Zone would be a difficult task that would involve fighting through compounds and prepared defences.

On 29 June, Hermon, Sergeant Ross, Corporal Huggett, Perry and Guardsman Barnett left PB Inkerman with roughly a whole platoon of ANA soldiers. The Royal Anglians pushed towards the Taliban stronghold from the north and the south while Hermon's ANA troops pushed in from the east, forcing the enemy back towards the river. Moore and De-Vall together with Lance Corporal Jackson remained in the PB in order to provide support from the heavy weapons. The Taliban were quick to realise that the British and Afghan forces intended to close in on them. Hermon's group came under fire less than a kilometre from the PB. As they were crossing an exposed poppy field, the Taliban opened up with a very heavy rate of machine gun fire. The British and ANA troops were pinned down by the fire which scythed across the fields cutting poppy stalks that fell around the prone soldiers. It was clear to Hermon that they would be unable to move forward and that they had to get out of the killing zone. He ordered the callsign to move back to the safety of some irrigation ditches only a few hundred metres from the PB. As they withdrew, Moore reported that the enemy were once again moving to try and flank them. Fortunately air support was on station for the operation and Hermon called in a US B-1 bomber which deposited a 1,000lb bomb onto the enemy positions. The Taliban were temporarily neutralised and after a short period of reorganisation the OMLT men were able to move west with their Afghan colleagues once again.

This first enemy ambush had come early and the Grenadiers wondered if the Taliban would defend the ground this hard all the way to the river. It didn't take long for them to find out. The advancing troops were ambushed twice more before they were able to link up with the Anglians close to the Helmand River. The Taliban sited their machine guns well and used their knowledge of the ground to best effect. Each of the ambushes was short but ferocious; the enemy attempted to withdraw after a period of heavy fire and before the British could fix their positions to strike them with aerial bombs or artillery. Hermon and his men led counter-assaults to try and cut the enemy off, but the Taliban managed to slip away. After one of these counter-assaults, an Apache was alerted to the enemy's likely escape route. The chopper quickly identified the fleeing Taliban fighters and was able to kill them before they reached safety.

The OMLT men were by now exhausted, they had only advanced a couple of kilometres into the Green Zone but had been forced to fight for most of the ground they had covered. When not in contact with the enemy they had moved cautiously, using the ground to best effect in case they came under fire once again. The threat from booby traps and IEDs was also on everyone's mind so they avoided obvious crossing points and vulnerable areas. Hermon was greatly relieved when they finally linked up with the Anglians in the late afternoon. It was good to know that there were large numbers of British troops nearby. The advancing troops now went into a hasty defensive position near the river while the next phase of the operation could be coordinated. The sense of relief was short-lived as the Taliban once again attacked the British and ANA troops as they reorganised themselves. A heavy rate of accurate fire was directed at the troops defending the hastily occupied positions. RPG rockets also exploded all around. In the initial salvo a Royal Anglian soldier was hit in the chest but his life was saved by his body armour. An RPG round exploded among a group of ANA soldiers, killing one instantly and seriously injuring two others. As fire was returned at the attacking Taliban, Barnett made his way to the badly injured Afghan soldiers. He did what he could to keep them alive but it was clear their injuries were severe. When the shooting eventually subsided the casualties were moved to a landing site where they were lifted to hospital. One of the two wounded soldiers died in hospital, but at least one of them survived due to the effective first aid of Barnett and the efficient evacuation system. It had been a very tough day ending in a body blow to OMLT morale. Hermon was devastated that he had lost two men, but he had to concentrate on the living.

The operation continued for several more days but although plenty of contact with the enemy was made, the fighting was not as intense for Hermon and his men. Most of the time was spent conducting aggressive patrols during the day and resting up in hastily occupied and defended compounds during the night. There was great relief when the operation finally ended and the ANA troops together with their mentors were withdrawn to PB Inkerman. The operation had been a great success; the Taliban had been caught in the trap and hundreds of enemy fighters had been killed or wounded. The ANA troops had been used well in support of the British infantry and the enemy had been cleared out of the area, for the time being anyway. It was difficult for the OMLT troops to return to their base with three men fewer than they had started with but they took solace from the fact that they had helped to smash the Taliban in the area. After Operation Ghartse Ghar the fighting in the area calmed down, unsurprisingly. It had been an intense period of operations in the Sangin area for the men of the Inkerman Company and they had been fortunate to have taken only one Grenadier casualty. The ANA had not been so fortunate. Their daily patrols among the population had come at a heavy price. By early July Ed Janvrin noted that 3rd Kandak had lost 13 men and had suffered another 13 wounded.

The month of June still had a sting in its tail for the Inkerman Company. The small OMLT detachment in Kajaki continued to work hard to restrict the Taliban near the little outpost at the northern extent of ISAF influence in Helmand. Frequent foot patrols were mounted with the ANA in the deserted villages and fields of Kajaki. The Grenadiers also led their Afghan colleagues through a series of localised operations with C Company of the Royal Anglians. On 30 June Lieutenant James Harrison led his OMLT troops on another of these joint operations. The weary British troops were roused from their sleeping bags at 0330 hours to commence their final preparations. After gathering their Afghan charges, the mixed patrol left through the rear exit of the little FOB and headed south. There were five other British soldiers with Harrison that morning: Sergeant Hughes, the platoon sergeant, Corporal Foy the medic, Lance Corporal Jordan, Guardsman Scanlon and Guardsman Dan Malcangi. They moved cautiously and peered through the early morning light for signs of enemy activity. A little way down Route 611 Harrison ordered the patrol to 'go firm' and to take up positions where they could observe and give support to the Royal Anglian callsign moving on their right. The patrol sniper immediately took up a position on a compound roof from where he had a good field of fire. He was good at what he did and had already accounted for a good number of enemy fighters in the area. As he peered over the sights of his sniper rifle the young soldier commented to Foy on how quiet the area was. Soldiers sometimes possess a sixth sense which tells them something is wrong. The two soldiers were just discussing why they felt things weren't right when Harrison ordered the platoon to move. He had been tasked to push further south and to take up a new overwatch position. From this new location the Afghan platoon could cover the next move by the Anglians through an area of the Green Zone known to the troops as 'Tally Alley'.

The OMLT men guided their Afghan colleagues along a dusty track which passed through the middle of the abandoned settlement. On each side there were high compound walls separated by narrow alleyways. Each of the compounds had a gateway or entrance but the mud walls were solid and straight, offering no cover. The patrol used both sides of the track and was staggered. Harrison was to the front left of the patrol and Guardsman Scanlon was at the front on the right side. Malcangi followed behind Scanlon with the Afghans dispersed among the other British troops. The lead part of the patrol was now sandwiched between two high walls either side of the track; for around 60 metres, there was no cover at all. It was an ideal ambush position and the Grenadiers were very uncomfortable. Suddenly the familiar sound of machine gun fire broke the quiet. Dust clouds sprang up along the track and bullets ricocheted off the mud walls leaving deep holes. Everyone dived for whatever protection they could find but, at the front of the patrol, cover was in short supply. Malcangi dropped to the ground and tried to make sense of what was happening. They were being attacked from the front and slightly to the right. It was a terrifying position to be in and he wondered whether to go forward or back. At that stage he heard a cry of pain coming from in front of him: Scanlon was down. Malcangi quickly half-crawled, half-ran to the injured soldier. Bullets were cracking overhead and striking the ground far too close for comfort. He saw no obvious signs of injury and thought that perhaps Scanlon had injured his leg. It was only when the wounded man announced that he had been shot and Malcangi rolled him over to locate the injury, that the wound was found. Malcangi realised the bullet had gone through the front of Scanlon's pelvis and had exited through his buttock. As he screamed into his radio and tried to get the attention of his platoon sergeant, he noted that the enemy were firing through loopholes in a compound wall about 150 metres away. Harrison was returning fire from the prone position on the left as rounds struck the dirt around him.

Malcangi realised that he would have to get Scanlon quickly into cover or neither of them would survive. From behind the wounded man, Malcangi gripped the Guardsman's body armour and pulled hard to drag him clear. To his annoyance the shoulder fastenings broke open and he was left with nowhere to grip. Scanlon was weighed down with equipment and was very heavy. Malcangi stripped away as much unnecessary equipment as he could and started to haul the bleeding man backwards. It was exhausting work and before long he found himself in a sitting position behind, dragging Scanlon along and then shuffling backwards for the next heave. After what seemed like an age the young Guardsman looked up to see Hughes and Foy bending over him. Between them they managed to haul Scanlon back into the relative safety of a compound entrance. The medic now set to work to stem the blood seeping from the exit wound in Scanlon's buttock and Harrison did his best to get the rest of the platoon into some decent cover. It wasn't long before Scanlon was successfully extracted, but the ordeal was not yet over for the remaining troops.

The Taliban now brought some accurate mortar fire to bear and the Grenadiers crouched against the compound walls for cover as the loud explosions shook the nearby ground. The enemy eventually withdrew and no further casualties were sustained. It was another narrow escape and Scanlon was fortunate that his injury was not too serious. The bullet had miraculously passed through his pelvis without causing critical damage. He had been saved by the actions of Malcangi, as well as Hughes and Foy who had run from their covered location to help Malcangi with the rescue.

Amazingly, the Taliban mounted an ambush from the same location only days later. On this occasion they were not as lucky. Coalition aircraft destroyed their positions in the compound and the OMLT men were able to follow up and to finish the enemy at the point of the bayonet. Several Taliban were killed and Hughes had the satisfaction of seeing Scanlon's attackers punished.

13

SUMMER HEAT

With the Inkerman Company dispersed around Sangin, 3 Company still fighting daily battles with the Taliban in Garmsir and the Queen's Company still spread between the patrol bases around Gereshk, the Grenadiers were being kept occupied. The men detached to the Anglians had been equally busy and Drummer Wintle had been evacuated after sustaining a back injury following an explosion. The other Grenadier casualties, including Guardsman Alex Harrison, were now recovering well in hospital at home in England. The men with the Londons too had seen their fair share of action and had done good work in the Gereshk area. The support and headquarters elements of the battalion were either already dispersed between the deployed companies or were working hard to ensure that the troops were sustained in the field. Small groups and individuals sometimes transited through Camp Shorabak for short periods of time and these were great opportunities to catch up. The returning warriors were usually bearded, tired and stank to high heaven. Their appearance was in stark contrast to the men stationed in Shorabak itself who lived in comparative luxury. All things were relative in Afghanistan and the troops who resided at the huge ISAF base in Kandahar were often held in contempt because of the perceived perks available in what was effectively a secure military town. A simple shower or a fresh meal was heaven to someone who had spent weeks or even months in one of the bases.

2 Company had been recuperating in Camp Shorabak where they conducted some low-level training with 2nd Kandak which was preparing to depart on a very hard-earned period of leave. Captain James Shaw was sent to Lashkar Gah where he met with the commander of Battlegroup South, which was based upon the Light Dragoons. Shaw received a preliminary briefing for Operation Bataka which was to commence at Garmsir in the near future. A strong force of ANA troops was required to take part in the operation and 2 Company would need to ensure that they were well-led and motivated. Shaw was told that the likely task for the Afghans would be to take over the dangerous Garmsir checkpoints from 3 Company. This would free up the Grenadiers to act as a manoeuvre force; after all, they knew the area better than anyone. The only Afghan troops available were the men of 2nd Kandak who were expecting to go on leave any day. They were understandably upset and disappointed when the news of their forthcoming mission was broken to them. Based upon the reports that had filtered through from Garmsir, 2 Company expected there to be fierce resistance from the enemy. Shaw was to command the OMLT group and he immediately asked for CSM Darren Westlake to come along as the warrant officer's experience would be invaluable. Shaw and Westlake put together a strong 12-man OMLT team that contained plenty of expertise. Lance Sergeant Tom Loder from the Anti-Tanks would provide the vital Javelin support, Lance Sergeant Matt Robinson, a mortar fire controller, would be on hand and T would also come along with his sniper rifle. Lieutenant Rupert Stevens was to command an ANA platoon and Guardsman Cheetham would act as the team medic. Shaw was also pleased to have the reliable Sergeant Byrne on board.

The mission required about 120 ANA soldiers – about a third of the kandak – and there was much deliberation among the Afghan commanders about who should be sent. They complained

ABOVE: Patrolling the dangerous Sangin roads. The view from the gunner's position standing in the rear of a WMIK with 7.62mm GPMG. Several vehicles were destroyed on this hazardous stretch.

BELOW: A deadly landmine.

ABOVE: The battered exterior of Sangin DC. It was from here that the isolated British infantry company, together with the Grenadier Guards liaison teams and their ANA colleagues patrolled to keep the Taliban out of the town.

BELOW: Small perks. The cooling waters running through Sangin DC gave some respite from the intense heat.

BOTTOM: Driving through Sangin on the infamous Route 611.

ABOVE: A view of FOB Delhi, Garmsir. The defensive sangars can be seen at the corners of the base in front of the smoke.

BELOW: Grenadiers on JTAC Hill, Garmsir. This position provided excellent views of the enemy occupied areas to the south but attracted much fire in return.

ABOVE: Catching up on some sleep, JTAC Hill.

BELOW: A closer view of the defences at JTAC Hill, Garmsir. Sergeant Betts is in the foreground.

RIGHT: On the roof at FOB Inkerman during a lull in battle. The large weapon on the tripod is the 40mm grenade machine gun. Javelin missiles can be seen in the foreground.

BELOW: The result of an enemy SPG 9 strike from outside FOB Inkerman. L to R: Lance Corporal 'Pez' Perry, Corporal Justin Huggett (Royal Australian Regiment) and Sergeant Dan Moore.

LEFT: Sergeant 'Tosh' De-Vall fires a javelin missile from FOB Inkerman. The javelin was a lethal weapon system highly rated by the troops.

ABOVE: 3 Company fighting patrol, Garmsir. Suppressive fire is brought to bear on the enemy as other Grenadiers assault. The man standing fires a 5.56mm light machine gun and in the foreground the heavier 7.62mm GPMG is used to good effect.

BELOW: Grenadiers and ANA troops clear a compound after an airstrike.

TOP: The isolated FOB Arnhem. Miles from anywhere it was very challenging to defend.

ABOVE: Troops run for cover during a rocket attack on FOB Arnhem.

RIGHT: Guardsman Lyne-Perkis demonstrates good spirits after being injured in a friendly fire incident with 2 Company.

LEFT: HRH Duke of Edinburgh, Colonel Grenadier Guards, meets Lance Sergeant Ball on a visit to the 1st Battalion in November 2007 while RSM Keeley looks on.

RIGHT: Ball in rehabilitation. He tragically lost a leg in an IED strike close to the end of the tour.

bitterly about the untimely nature of the task and felt very hard done by. There was little that the OMLT troops could do about the situation, the mission would go ahead whatever happened and the Afghans reluctantly accepted the fact that they would be deploying on another hazardous operation. The 2nd Kandak commanders were briefed on the nature of the mission and Shaw was assured that the nominated troops would be ready to deploy according to the orders they had been given. He double-checked the timings with the Afghan commander.

Together with their Grenadier mentors, 2nd Kandak were to drive 120km across the desert to Garmsir. The departure was planned for the early morning on 15 June. A huge amount of preparation had been done by Shaw and the OMLT men and they were ready for a 0500 hours start in order to cover some ground before the heat of the day made life unbearable. When they arrived at the kandak HQ in Camp Shorabak, there was, alarmingly, no sign of life. It took some time to track down the kandak officers, who were still in bed. There had been a major failure in communications and the ANA men were subsequently not ready to move until several hours later. By now the sun was high in the sky, the heat was appalling and it was far from the start that Shaw had anticipated.

The tired Grenadier and disgruntled ANA convoy eventually left Shorabak and headed south into the desert. The first few kilometres passed uneventfully with Westlake leading the convoy. There were three OMLT vehicles, eight Ford Rangers and four of the heavy 6 x 6 trucks with their notoriously poor cross-country capability. True to form, as soon as the convoy started to cross the wide wadis, the bottoms of which consisted of soft sand, the trucks became stuck. Much time was lost in towing the stranded vehicles out of the dry river beds. Westlake carefully selected firm routes so that the heavy trucks avoided the soft sand, but even this proved to be problematic. Things were now becoming so bad that it was impossible to cover more than 500 metres without one of the trucks becoming stuck. The tough terrain took its toll on the vehicles and by now several of them were being towed. After nine hours of utter frustration the convoy had covered only 14km of the 120km journey and they were still in sight of Camp Shorabak. Shaw was completely exasperated by the situation and reluctantly made the decision to turn back. Several hours later the convoy arrived back at its start point, angry, tired and frustrated. Shaw arranged for the convoy to reorganise itself and he briefed the commanders that they would try again in the morning.

The next day further communication and administrative difficulties delayed the convoy's departure until around 0700 hours. Shaw was now very concerned; they were already a day late and the operation hinged on the ANA taking over the checkpoints from 3 Company. On the up side the ANA seemed to have found some competent drivers and the first 15km was covered without any significant difficulty. Their luck couldn't last and one of the ANA trucks suffered catastrophic damage in a particularly deep wadi. Shaw's heart sank; the dripping oil was indication enough that this particular truck would be going no further. The Grenadiers fumed and Shaw contacted Captain James Greaves, the operations officer, by radio. He arranged for a replacement truck to be sent out, but warned Greaves that they should prepare for failure and that they might have to examine an alternative option for the operation. The replacement truck duly arrived and the stores were cross-loaded before Shaw once again headed south with his rag-tag convoy.

Shaw couldn't help worrying that the convoy would become stuck further away from help. But his luck seemed to be holding and the convoy found itself approaching FOB Dwyer at around 1730 hours. They were almost two days late but they had made it. At the sight of their destination the Afghans suddenly came alive. For the last 500 metres the horns of the vehicles were depressed, making a terrible racket. Green, red and black Afghan flags were produced and the passengers cheered loudly as they drove into the FOB. It was as if a liberating army had arrived, at least in the minds of the ANA, who were enthusiastically greeted by their colleagues in the FOB. The British troops looked on and wondered what on earth was happening. The OMLT men jumped down from their vehicles, grateful that they had finally arrived. They were better late than never and the Afghans were showing enthusiasm. Shaw was ushered straight into the battlegroup commander's orders group and Westlake organised the administration. The area that the new arrivals had been allocated to set up camp was only about 50 metres from the newly reinforced 105mm gun positions and Westlake hoped that they wouldn't be firing on this particular evening. He was to be disappointed and the guns fired a number of rounds in the early hours, making it impossible to get much sleep.

The following morning they were able to see just how isolated and barren the FOB was. Just a line of bastion walls marking the perimeter and little else. Outside the walls there was nothing but desert as far as the eye could see. It was incredibly hot and there was very little shade. At 1200 hours the orders group assembled once again. Each of the commanders described his part in the operation in turn so that everyone understood not only their own but also everyone else's role in the mission

Shaw's ANA troops were to relieve 3 Company who would reorganise in FOB Delhi where a battery of light guns had been especially flown in to provide a heavy bombardment before the operation. The guns would mask the movement by British troops around the base. At H-hour, A Company of 1 WFR would cross the canal using the light infantry bridges. They were to secure the far bank which would allow the engineers to build a bridge over the canal in the area known as Balaclava. This would be the first time that the engineers had built a major bridge under fire for some decades. 3 Company would then cross over and move forward to clear a series of enemy positions which had been troublesome for some time. This would be the most dangerous part of the operation but the Grenadiers were eager to get to grips with the enemy and to gain some 'pay back' for Guardsmen Simon Davison and Daniel Probyn. Apache helicopters would be on station and coalition aircraft would be on call if required.

The BRF were, as usual, to be heavily involved in this latest operation and they too had arrived in Garmsir on 16 June. Their role was to operate behind the forward line of the enemy troops, pushing south and east, close to a bend in the Helmand River known as the 'Fishhook'. Once in place they were to locate any enemy positions and disrupt them as much as possible. This was all designed to take some of the pressure off Garmsir and to make movement at the enemy rear difficult. The BRF element was named Operation Anthracite and they knew that it would be a dangerous undertaking. They were operating almost at the limit of helicopter support range and there was nothing to the south but the enemy.

On the afternoon of 17 June, Shaw led his ANA troops across the desert to FOB Delhi and soon hit soft sand. Shaw's heart sank; after only a few hundred metres they were stuck once again. They persevered and eventually managed to extract themselves but as they neared Garmsir the OMLT men saw another large area of soft sand through which they would have to travel. Worse still this was a key vulnerable point and they feared coming under enemy fire if they became stuck. Luckily, fortune smiled on the little convoy and they made it safely through to FOB Delhi where they were met by the familiar faces of 3 Company, although those faces were now mostly bearded. Later that day, Shaw, with two platoons of ANA troops, took over Balaclava and Byrne took a third platoon to relieve the Grenadiers at JTAC Hill. The OMLT commanders did their best to settle the ANA troops into the little bastions, but there were far more of them than there was available accommodation and they were rather cramped. However, before long the sound of jolly Afghan music from transistor radios was filling the air. Shots rang out as a soldier accidentally fired his weapon, but Shaw was by now far from surprised by this kind of thing.

The following day General Muhayadin Gori arrived in Garmsir to see the situation for himself. He toured the positions and chatted to the Afghan police as well as to his own men. He looked very proud that his troops were now garrisoning Garmsir, which until recently had been a Taliban stronghold. The significance of so many Afghan soldiers arriving in the town was not lost on the civilian population either. By 18 June the OMLT and ANA troops were settling in to a decent routine and Shaw was starting to relax. That afternoon the Royal Artillery gunners in FOB Dwyer adjusted their guns for the operation. The Grenadiers in FOB Delhi and in the checkpoints watched the barrage from behind their bastion walls. No one had seen anything quite like it. In the distance they saw the orange flashes as the high explosive shells landed. Seconds later the rumble of the detonations reached their ears. It was a truly impressive sight and the troops looked on in awe of the explosive power.

Later that evening Shaw's party spotted movement about 1 km away in a deserted hamlet. Through their night sights they could see the faint glow of torches. Intelligence suggested that the enemy were searching for the dead and wounded from the earlier artillery barrage. Westlake immediately opened up with the GMG which was quickly joined by the other weapons systems available. Robinson called in an artillery fire mission and after a few adjustments all six of the 105mm guns from FOB Dwyer opened up. The effect was awesome. Within a minute, 72 high explosive shells fell on the target. There was no more movement in the area that night.

Operation Bataka was due to start the following evening, 19 June, but was postponed due to a shortage of helicopters. The Taliban were unnervingly quiet but this was probably due to the artillery pounding they had received the day before and the troop build up in Garmsir, which could hardly have gone unnoticed. The summer heat was now torturous and Shaw recorded the temperature at the eastern checkpoint as being 52°C; in direct sunlight it was as high as 63°C. The Grenadiers sought the shade of their vehicles to escape the intense heat when they were not observing for enemy activity. The ANA troops seemed to have a knack for finding shady corners; they were used to the extreme heat and wasted no energy.

The BRF had spent the previous couple of days to the south of Garmsir where they had conducted some reconnaissance to try and identify areas where the Taliban were concentrated. They had been briefed to disrupt enemy operations to the south in order to make life more difficult for the Taliban around Garmsir. It was now decided that they would probe into an area of the Green Zone known as the 'Y junction'. By causing confusion and disruption in the enemy rear, the reconnaissance troops would make it much more difficult for the Taliban to resupply or reinforce the fighters dug in to the north.

The day before the start of Operation Bataka, Captain Piers Ashfield gave orders to 1 Platoon for a dismounted advance to contact into the Green Zone. Ashfield was an experienced officer who had seen action with the Guards Parachute Platoon with 3 Para the previous summer; he had recently returned to Afghanistan and was thrown in the deep end with the BRF. At first light the following day 1 Platoon stepped off with one section leading and two further back. The platoon sergeant, Howard Lawn, located himself towards the rear and watched with awe through the early morning light as his platoon deployed in conventional infantry formations as though they were training in the Brecon Beacons. 2 Platoon remained mounted in order to give fire support if it was required. Ashfield directed his lead section from the centre of the formation and before long they had pushed about a kilometre into the Green Zone. Local intelligence reported by the interpreters suggested that the Taliban knew that the British were in the area but clearly did not yet know where. It was not usual to encounter coalition troops this far south and already confusion was being sown in the minds of the enemy. Before long the platoon had pushed up to a dirt road running along a canal bank and it was decided to set a snap ambush to see what happened. It was clear that the enemy did not know exactly where the British were. As the BRF troops settled into the available cover, 1 Section reported a white pickup truck moving towards them at some speed. Ashfield instructed the section commander, Sergeant Andy Austin, to stop and search the truck. Austin moved into the road and as the vehicle sped towards him he indicated for the truck to stop. There was no response and a warning shot was fired. This had the desired effect and the truck slowed to a halt in a cloud of dust. Austin noticed that there were at least three occupants and the men of his section shouted for the men to dismount. As they were doing so the platoon came under contact from their front. RPG and automatic fire could be heard and before long this intensified.

Lawn moved up to support Austin's section and discovered that the three men who had been detained from the pickup truck all had webbing and AK-47 rifles. They also had portable radios and were obviously Taliban fighters. This was a result: three prisoners who would yield valuable intelligence. The enemy fire was now becoming quite heavy and Lawn marshalled his prisoners to the rear so that they could be processed and eventually sent back to the CSM. At the same time, he tried to locate the enemy firing points. There was a suspicion that the enemy was already attempting to get behind them.

As all this was taking place, Lawn heard a shout of 'man down' over his radio. It was Lance Sergeant Dougherty, second in command of 3 Section. Lawn quickly made his way over where he found that Sergeant Dan Aldous had been shot in the chest. The wound was on the left side, just to the right of his body armour plate. It was clear that the wound was serious and that Aldous would need to be evacuated rapidly. Lance Corporal Melville, the medic, quickly went to work on the wounded soldier and got him stabilised. The firefight had intensified and things were becoming fairly unhealthy. CSM Ian Farrell once more moved forward to the pinned-down group and helped to organise the evacuation of Aldous back to the emergency helicopter landing site, which was some 1,500 metres back into the open desert. Lawn had already sent a casualty evacuation request and the MERT was taking off from the distant Camp Bastion. The prisoners were also sent rearwards with an escort.

Ashfield, Lawn and the other men in the point platoon now concentrated on destroying the enemy that had engaged them. A fire mission was requested from the 105mm guns at FOB Dwyer. An initial spotting round was fired to ensure that the adjustment was correct and then five rounds of fire was requested. The BRF men expected the artillery to fall accurately on the enemy positions and they waited eagerly for the sound of the shells passing overhead. The first round impacted just short of Ashfield's own position with a deafening crump and a huge dust cloud. Something was horribly wrong and Lawn could only look on as the remaining four rounds landed almost on top of the platoon commander's position. Lawn's heart was in his mouth. He couldn't see how anyone could have survived the explosions and he wondered how the hell he would be able to evacuate the casualties. The artillery was ordered to check fire and Lawn rushed to his platoon commander's aid. On arrival he was shocked and relieved to find Ashfield cursing loudly and brushing a thick coating of dust from his clothes. Miraculously, no one had been injured, although the antenna for their communications equipment had not been so lucky.

With the help of an American B-1 bomber and support from 2 Platoon, 1 Platoon was able to safely extract themselves and to rejoin their vehicles. The CSM told Lawn that Aldous had been evacuated, and that he was stable and in good heart when he was lifted out. He had even donated his cigarettes to Lawn before departing, a gesture that reassured everyone as to his condition. Farrell and Colour Sergeant Frith were again engaged by the enemy while moving about and once more called upon the US Air Force for support. More bombs were dropped and the enemy fire was silenced. The BRF moved out into the open desert where they leaguered up for the night. Their three detainees were still with them and after some tactical questioning about further Taliban positions in the area they were lifted out by helicopter. It was an eventful first day and the BRF had more to do in support of Operation Bataka.

Their next target was an area where two main routes joined together and was known as 'Sierra'. Intelligence suggested that this was a holding area for the Taliban and that there were other facilities in the area. Covert surveillance was conducted before it was decided to make a night approach when the enemy were least alert. Unfortunately, things did not go to plan; on the way in one of the vehicles developed a serious mechanical fault and made a loud screeching noise. This could clearly be heard for miles and sure enough it attracted a heavy rate of fire from the objective area. Luckily the Taliban were firing blind and had got the direction completely wrong, directing fire about 1km to the south. They continued their advance on foot according to their original plan and soon approached the area of Sierra. As they drew nearer they started to encounter groups of civilians fleeing the area, also convinced that the British were coming from the area the Taliban were engaging.

It didn't take long for the enemy to realise their error and the BRF lead teams soon found themselves under intense small arms and RPG fire. The fire became so heavy that the lead elements decided to withdraw to a safer area. During the move back, two of the team became separated and were pinned down by accurate fire. They were in danger of being overrun and for a while the situation was desperate. Repeated attempts were made to fight through to the stranded soldiers, but these were all beaten back. A well-aimed Javelin missile caused sufficient shock and confusion to stun the enemy for long enough for the trapped soldiers to fight their way back to safety. Air support was on station quickly and the BRF were able to withdraw under the cover of a couple of 500lb bombs that were dropped onto the Taliban firing points. Even after contact was broken, the Taliban followed up and it was necessary to deal with another group firing from a graveyard. The little convoy of vehicles escaped into the desert expanse without taking any casualties and knowing that they had again caused the Taliban some real problems. The areas that had been targeted in the last few days had previously been safe areas for the insurgents and a good deal of disruption had been caused.

To the north, in the patrol bases above Gereshk, the Queen's Company were still widely dispersed in their small OMLT teams. Their routine was largely unchanged. They mounted daily patrols to secure the area and deter the enemy from interfering with the local population. Their presence was a useful buffer to keep the Taliban out of Gereshk. Since they deployed on Operation Silicon they had been in this dangerous area without a break for around nine weeks. The enemy had shown their faces regularly and there had been a number of small engagements. The Taliban were still intent on laying their deadly IEDs and the tracks around the bases and to the south were extremely hazardous, especially on the tow path leading to Gereshk. 1 WFR had lost a man in an action near the northern PB and the ANA commander of 1 Company and one of his platoon commanders were both killed instantly in a powerful explosion at the beginning of June. There were several near misses and a number of devices were found before they had a chance to harm anyone else.

On 20 June the company was consolidated in the area and preparing for an operation. Second Lieutenant Folarin Kuku led his small OMLT team on a foot patrol between PB South and PB Centre. A short distance behind him was Lance Corporal Jack Mizon and across the road Guardsman name Barnes moved in an alert posture. The patrol had moved about 400 metres along the track when Mizon saw a couple of local men with a motorbike about 150 metres ahead. He was initially suspicious, but the men looked relaxed and unconcerned so he turned his attention elsewhere. The patrol made its way along the dusty track but its progress was halted by a huge explosion. Those who were not thrown to the ground quickly dived for cover. Mizon found himself on the ground slightly dazed with a loud ringing in his ears. The area ahead was shrouded in a huge dust cloud and he could see little else. As he started to regain his hearing Mizon made out the sound of someone shouting for help. He pulled himself to his feet and ran through the dust cloud until he found Kuku lying at the side of the road. He quickly knelt by the young officer's side. Still shocked from the blast, he now desperately tried to make an assessment of his platoon commander's injuries. Kuku was conscious and was able to speak to Mizon which helped to calm the young NCO. Kuku's combat trousers had been shredded and blown away and some serious shrapnel injuries could be seen on his legs. Mizon was joined by Barnes and Lance Sergeant McDonald and the trio applied field dressings to the officer's injuries. Kuku's interpreter had also been injured and he too received medical attention from the worried OMLT troops. Lance Sergeant Dragon used his radio to inform Major Martin David and the company HQ about what had happened. David immediately sent a request for a casualty evacuation and CSM Glenn Snazle set out to locate and clear a suitable landing site for the helicopter.

Meanwhile, Lieutenant Paddy Hennessey used his vehicles to speed to the aid of his injured friend. Kuku was loaded onto the WMIK which drove to the landing site. Snazle looked over the casualty and concluded that his men had done a good job of patching him up. Unfortunately, Kuku was virtually naked from the waist down and Snazle did his best to cover the young officer's modesty with a shemagh. More bad news now filtered through. The MERT helicopter was already dealing with a casualty elsewhere and another routine flight would have to be diverted to collect the wounded officer.

Some miles away Captain Rob Worthington was aboard a packed Chinook. He was being conveyed back to Bastion for his R&R along with Sergeant Davis and a number of other troops. They were told that the aircraft was being diverted and noted a sudden change of direction. In a matter of minutes Worthington recognised the area that he had left very recently. Davis and Worthington were horrified when they saw two casualties being carried towards the vibrating helicopter and recognised the outlines of the tall Queen's Company men running towards the ramp. Davis found himself looking at his own platoon commander, with whom he had parted only hours before, lying bloody on the floor. A number of civilian media types were also on the helicopter and they looked on wide-eyed. The RAF pilot wasted no time and lifted off for Bastion. Kuku was comforted the whole way by his own platoon sergeant and by Worthington. Kuku's injuries were serious but fortunately not life-threatening and he was quickly stabilised.

Meanwhile, as the Queen's Company men watched the Chinook disappear into the distance, Mizon, McDonald and the others dusted themselves off and wondered how Kuku had survived the huge blast. It was curious that relatively small amounts of explosives could be fatal, but occasionally physics contrived to make a blast move in a particular direction. The patrol had been very fortunate not to have lost anyone. Within a short period of time David was giving orders and the operation continued as planned.

Back in Garmsir all of the vital elements were in place for Operation Bataka to commence. Thursday 21 June was another unbearably hot day during which everyone sought the shade. Stevens deployed with a platoon of ANA troops to an Afghan police checkpoint known as 'Nijmegen'. This was about 2km north of Balaclava. They would deal with any Taliban activity coming in that direction as a result of the planned operation which was soon due to get under way. Loder and Robinson accompanied the ANA platoon; this ensured that mortar and Javelin support would be available if needed; once in place they dug in. During the afternoon final preparations and rehearsals were conducted for what would be a difficult operation conducted in darkness. The troops lay sweating in whatever shade they could find, trying to snatch some rest for the long night ahead of them. At 1940 hours the gunners at FOB Dwyer commenced the artillery barrage which would precede the bridge construction in the area of Balaclava by the engineers. The bombardment was intense and the 105mm shells could be heard whistling through the air before they struck their targets with a loud crump. Bright orange flashes illuminated the night sky and black smoke rose in the moonlight. Shell after shell burst on the enemy positions and it seemed as though nothing could survive under the awesome power of the artillery which was now joined by mortar fire from FOB Delhi. At 2000 hours, A Company of 1 WFR moved forward with their infantry bridges and, under cover of the barrage, crossed the canal to secure the far bank. Intelligence suggested increased activity by the Taliban but they made no effort to oppose the British troops who were now pushing forward to defend the bridgehead. Once the immediate area was secure the engineers moved forward to build the bridge that would remain in place over the canal. This was an incredibly noisy operation as the heavy trucks manoeuvred around the bridge site and the engineers moved the heavy steel bridge sections into place. The operation continued throughout the night and the bridge slowly took shape; still there was no sign of any enemy activity and their positions continued to be harassed by the British mortars.

At 0500 hours 3 Company moved forward for the final and most dangerous phase of the operation. They patrolled out from Delhi, bayonets fixed, with a steely determination to inflict the maximum damage on the Taliban fighters who had been responsible for the deaths of their comrades. The Grenadiers moved with practiced confidence to their final assault positions, using the now familiar aluminium infantry bridge to cross a steep-sided irrigation ditch close to the objective. 3 Platoon led the company and Sergeant Scott Roughley was again in the lead section. The objective consisted of around 25 compounds that squashed together forming a small hamlet. It was clear from the air photographs that the area was a maze of alleyways and rat runs. It was going to be very difficult. 3 Platoon was to break into the urban area through a wall by use of a bar mine. They reached the wall of the compound without being detected and Major Will Mace gave the command to start the attack. The bar mine was detonated and a huge shockwave ripped through the hamlet, raising a massive dust cloud. A large hole was blown in the compound wall and Roughley launched himself through the dust. As he emerged from the smoke inside the compound he tried to orientate himself and to his shock he suddenly saw a single Taliban fighter. The man was disentangling himself from a wicker basket which had been lowered from a tall tree on the edge of the compound. Roughley quickly dealt with the man and as he did so he was aware of shouting behind him. Lance Corporal Bocock had spotted two more enemy fighters in a bunker; they were too slow and were dealt with by use of grenades. Other desperate Taliban were seen clambering away across the rubble in the half-light, pursued by tracer which zipped off the buildings. The enemy must have been caught by surprise. When he considered the situation later, Roughley assessed that a sentry position had been constructed in the tall tree and that the sentry was hoisted up in the wicker basket by means of a pulley system. This sentry had clearly not been alert and his idleness had resulted in his own death.

The remaining platoons passed through the first compound held by 3 Platoon and the clearance continued as the Grenadiers bounced from compound to compound. As the Taliban recovered they returned fire, but the initiative was with the men of 3 Company who assaulted through their objectives, efficiently clearing the myriad of buildings and trench systems with rifles and grenades.

Back at Balaclava, Shaw followed the progress of the assault on the radio and by listening to the bursts of distant automatic fire punctuated by the crump of hand grenades. Inevitably the Taliban now realised that 3 Company intended to clear them from the area in a determined assault and they started to respond. Enemy fire was directed at the assaulting Grenadiers from several different directions as the mutually supporting enemy positions opened up. The Guards replied, firing 81mm mortars onto these positions as they were identified and the AH 64 Apaches that were on station for the operation now started strafing runs in support of the ground troops. The Apaches were lethal in this role and their hi-tech thermal imaging devices sought out the Taliban in their fire positions. Unfortunately for the men on the ground, a good deal of metal in the form of empty shell casings tended to fall from the helicopters onto the Grenadiers below. No one wanted to be taken out by one of the heavy metal casings and the Apache was told to move in case someone below was injured.

JTAC Hill now came under attack and Byrne responded by calling in both mortar fire and coalition aircraft to attack the enemy. Two 500lb bombs were dropped and the attack soon petered out. A number of Taliban were identified in a very deep ditch line; they had excellent cover from fire and were proving very difficult to kill. Two US A10 fighters were given the task of dealing with them and Shaw watched as the highly manoeuvrable ground attack planes swept in low over Balaclava. They dived down low and fired their cannon along the ditch line. There was a loud 'buuuurp' as the 20mm shells ripped along the ditch, throwing dust and shrapnel into the air. It was immediately clear that the strafing run was successful and loud cheering was heard from the British positions. As he monitored the radio, Shaw heard the US pilot report in a slow Texan drawl, 'They ain't going nowhere.'

The 3 Company assault was entirely successful and several enemy fighters were killed on the objectives. There were no friendly forces casualties and the Grenadiers were still up for a fight and wanted to ensure that the Taliban were pushed right back from the area. Unfortunately, to clear such an urban area would require many more men. Mace rightly assessed that the company was in danger of becoming too stretched and he ordered them to hold their current positions. The Grenadiers traded blows with the Taliban from these new positions without trying to clear any further compounds. They knew only too well that the enemy would be quick to move behind them given the opportunity.

After some hours in these positions, Mace was ordered to withdraw across the bridge and into FOB Delhi. Under the cover of mortars, 3 Company passed back across the bridge at Balaclava where they encountered many exhausted engineers who had worked throughout the night to build the now completed structure. From their positions around Balaclava, Shaw's Grenadiers watched with pride as their 3 Company comrades emerged from the smoke and mist, heading back towards FOB Delhi. They were grateful that no one had been killed or injured and their returning friends made an impressive sight, walking tall as confident, battle-hardened men. These 'Garmsir Grenadiers' had fought hard for several months and it was clear that they had relished the opportunity to take the battle to the Taliban in such an effective way. Operation Bataka had been a complete success; the Taliban, although not yet vanquished, had definitely taken a beating.

The following days saw the troops in Garmsir return to a routine and the various assets that had been put in place for the operation were steadily withdrawn. Shaw and his ANA troops returned to Shorabak without Stevens, who was now cross posted to 3 Company. The BRF returned to Camp Bastion, but on the way back they passed again through the area of the Y junction where on the first day of the operation Aldous had been wounded. In a small settlement close to where the fighting had taken place they found that the local civilians had been murdered or driven off by the Taliban who had accused them of being spies for ISAF. This was of course completely untrue; the BRF had had no contact with them before the action. It was another example of the ruthless idiocy of the Taliban. This was not the only nasty surprise waiting for the British at the Y junction. The Taliban had left a booby-trapped ammunition tin close by and this detonated a large IED as the BRF approached. No one was injured but it was a close call.

14

BLOODY JULY

The month of July began with some intense fighting to the north of Sangin as Operation Ghartse Ghar placed further pressure on the Taliban. Task Force Helmand was working hard to maintain the momentum and to keep forcing the enemy out of the towns and villages in order to further reduce their influence. Wherever concentrations of Taliban fighters were detected, they were targeted. It was impossible to track all of the small groups who infiltrated back into the cleared areas, but it was possible to attack the enemy strongholds from where their operations were planned and supported. Coalition surveillance assets were deployed to locate these strongholds and the BRF was frequently used to confirm the enemy positions.

The area of the Upper Gereshk valley between the three newly established patrol bases north of Gereshk and to the south of Sangin remained difficult. British patrols venturing north or east from the PBs frequently came into contact with the Taliban and the enemy still penetrated to the south to lay their deadly IEDs. Task force HQ had planned an operation to clear some of the enemy strongholds from this area of the valley. This operation was to be called Operation Tufaan and would be carried out by elements of Battlegroup Centre supported by 1st Kandak and their Queen's Company mentors. It was set to commence at the end of the first week in July and would overlap with others continuing in the Sangin area. It was also necessary for small patrols to transit to and from FOB Price in order to carry out various administrative tasks in order to sustain the dispersed company in the field and to ready themselves for the tough job ahead.

Tragedy hit the Grenadiers once more at the very beginning of July when a two vehicle patrol from the Queen's Company was ambushed in Gereshk. One man was killed and four others were seriously injured, one of them critically. The incident served as a reminder of just how vulnerable the troops were to this type of attack in their WMIKs. The incident served as a reminder of just how vulnerable the troops were to this type of attack in their WMIKs.

In the following days the Queen's Company was relieved in the Gereshk PBs by the men of 2 Company. Most had been deployed for more than two and a half months without a break and they looked forward to the relative comforts of the battalion's main base upon completion of this latest operation. The Queen's Company would remain under the operational control of Battlegroup Centre for Operation Tufaan. The commanding officer's intent for this operation was to move east of the Gereshk PBs to locate any enemy there. Once they were located, the battlegroup, including 1st Kandak, would strike in order to destroy them. This was expected to take place near the villages of Rahim-Kalay and Kakaran, a distance of about 5km further east of the bases, in the Green Zone. The objective of this operation was not simply to kill more Taliban. The troops would also look for opportunities to exploit their success and to influence the local population. Having defeated the enemy in the area of Rahim Kalay, a further PB was to be established so that a presence could be maintained in the area.

Before any major strike was made against the Taliban, efforts were made to make the enemy think that the British and Afghans intended to continue raiding the area rather than to mount a decisive operation designed to clear the vicinity. The BRF continued to screen the villages and the edges of the Green Zone where they probed for enemy positions. All available surveillance assets were used to locate the enemy in order that they could be destroyed in their strong points. Other elements conducted operations deep in the enemy rear to disrupt their command, control and lines of communication. By establishing further patrol bases in the area, a permanent security presence could be maintained and the Taliban could be deterred from influencing the local population. However, about a week before the operation was due to commence, the plan changed. It was decided that Tufaan was no longer the brigade main effort. This meant that a significant amount of assets allocated to the operation were cut away. The operation was now to be called Leg Tufaan which roughly translated meant Mini-Tufaan. The change of name was insignificant but the withdrawal of assets and the reduction in manpower was going to be problematic.

As the shaping phase of the operation unfolded, the BRF and other reconnaissance elements were able to identify detailed enemy positions and the plan was refined. Confirmatory orders were issued by the battlegroup commander. B Company of 1 WFR – now known as 2 Mercian after changes to the structure of the British Infantry – were to assault Rahim Kalay and, simultaneously, 1st Kandak along with their Grenadier mentors were to attack the area of Kakaran. A Czech special operations group would act as the battlegroup reserve and would be located near tB Company, which had men from the Londons and the Grenadiers attached. David delivered his orders to the Queen's Company on 5 July. H-hour was set for 0330 hours on 7 July. 1st Kandak, led by the Grenadier OMLT element, were to clear the area of Kakaran and then move to do the same in the village of Adin Zai. The orders were clear and everyone concerned understood the plan, but it was plain to see that the late changes had restricted the support available on the ground. The ANA and OMLT were to advance through the dangerous Green Zone on foot. There was no support to the south and the Queen's Company would be dependent on the progress of 2 Mercian to the north for security there. The terrain they were to operate in was extremely difficult; thick vegetation, irrigation ditches with limited crossing points and a myriad of compounds, each of which would have to be carefully cleared. The ground to the north was more open and vehicles were able to operate there. The Grenadier section commanders were quick to note that the task they had been given in the Green Zone was going to be very tough for them and that they were going to have to clear more compounds than the more capable mobile force to their north. CSM Glenn Snazle was particularly concerned that at points they would be more than 4km from the nearest vehicle access point in an area where the MERT helicopters would find it challenging to land. Any casualties would be very difficult to extract.

CSM Edgell was now in command of the wounded Folarin Kuku's platoon. Because of the recent casualties David had been forced to reorganise the company into two OMLT teams and a small HQ. The Queen's Company would be under-strength even before they started what looked to be a very difficult dismounted operation. On 6 July the men spent the day in rehearsals and administration before moving off from their bases to be in position for the 0330 hours start. The insertion of B Company and the 1st Kandak was more difficult than expected due to the exceptionally low light levels and movement was slow. There was further confusion when a Fire Support Team WMIK was damaged by an explosion; fortunately there were no casualties on this occasion but the noise no doubt alerted everyone in the area that ISAF forces were on the move. The line of departure was crossed at 0330 hours with Lieutenant Paddy Hennessey's Amber 61 leading the 1st Kandak with the remainder of the company following on. Snazle brought up the rear on his little quad bike which was loaded with ammunition and water. Both B Company and the ANA quickly gained footholds among the compounds of their objectives in Rahim Kalay and Karakan. Evidence of enemy activity was soon discovered. Tunnels and abandoned munitions indicated that the Taliban had been here very recently, and before long B Company was in contact in Rahim Kalay. Hennessey's ANA troops had successfully cleared the first compounds but a short time later they too were being engaged from enemy positions. It was only 0745 hours and the Taliban were demonstrating some fierce resistance. Effective small arms fire cracked overhead and was punctuated by the whoosh and bang of RPG rounds. The ANA gave as good as they got and the battle raged on as slow progress was made. Although it was still early morning the sun was already scorching and the going was tough for the heavily laden Queen's Company men who were operating without any vehicle support. The CSM had been forced by the terrain to abandon his quad bike; maintaining resupply was virtually impossible as the additional ammunition and water that the CSM's party had carried forward on foot was used up.

Where the enemy firing points were identified, 81mm mortars were used to silence them with the reassuring crump of high explosive. Dust and smoke surrounded the various compounds and made for a confusing picture. The Taliban were adept at quickly changing their positions and at exploiting any weakness shown by their adversaries. Movements were detected in the undergrowth and irrigation ditches as the enemy fighters tried to outflank the ANA. At one point Hennessey's ANA troops hesitated and then failed to follow him during an assault. Undeterred, Hennessey, with only three other Grenadiers, successfully continued the attack, killing several Taliban in the process. Seeing the Grenadiers assault a pre-prepared enemy position and come out on top, the ANA troops were rejuvenated and rejoined the battle.

Apaches were soon on the scene, their 30mm cannons ripping into the compounds. As the Taliban withdrew from their positions they were quickly occupied by the OMLT and ANA troops who brought fire onto the next enemy position. The Taliban fought doggedly, ensuring that the close quarter battle continued all morning, with the Apaches above proving to be the decisive force that often halted enemy counter-attacks. At one point, one of the

Apaches engaged the compound occupied by callsign Amber 63, firing a number of bursts and a Hellfire missile at the collection of little buildings. This was an unpleasant surprise for the Grenadiers who immediately called for a 'check fire'. The ANA claimed to have had men killed in the attack and David tried desperately to ascertain an accurate picture from his position in one of the newly occupied compounds. Fortunately no one had been killed or seriously injured, but it had been a very close call. David quickly realised that the spot maps were inaccurate and the men must immediately stop relying on them; when fire was being called in this close, a mere 80 metres could prove critical.

As the battle raged on, David moved up to the position held by Amber 63. Heat was becoming a major problem for the exhausted Grenadiers and water was in very short supply. There was a brief respite when, during a lull in battle, it was assessed that the enemy had withdrawn to reorganise. David decided to consolidate his forces on the objectives already seized. A number of standing patrols and observation posts were established in order to monitor enemy activity. While the OMLT paused for breath, an emergency resupply of water and ammunition was requested but turned down. They had been on the move for nine and a half hours and had been fighting for more than six of those. In this sort of heat and intense physical activity a soldier probably needs to take on in excess of ten litres of water per day. There was no way that this amount of water could be carried on foot. The enemy continued to probe the positions and gun battles continued around the edges of the objective as the Taliban attempted to encircle the little ISAF force. The heat of the mid-day sun was now unbearable for men weighed down by heavy body armour and helmets. It was plain to see that many of the men were worn out and some were displaying the signs of heat exhaustion, a serious and potentially life-threatening condition. The medics recorded body temperatures of 103 degrees on a couple of the soldiers; they needed urgent medical attention and the MERT was requested.

With the enemy so close Snazle struggled to identify a safe heli-copter landing site from which the casualties could be extracted, making it necessary to move the exhausted men more than 800 metres across difficult terrain. This in itself was a mammoth task. Sergeant Alexander, on the roof of a compound, kept the enemys' heads down with his GPMG group and coordinated with Hennessey, who moved a small group out to ensure that the enemy did not interfere with the extraction. Snazle then broke out with a small section carrying the now very serious casualties. This was exhausting work in the heat and men stumbled and fell, worn out under the weight of their equipment, before hauling themselves up again. The enemy continued to probe and to try and surround the OMLT troops. Lieutenant Will Harries and his team fought a particularly savage battle against a determined group of enemy as close as 100 metres from them, while Hennessey's group, too, neutralised the Taliban who were probing their flank. Harries and a small group of his men pressed their faces into the dirt in a small ditch as the air support dropped bombs onto a firing point in a compound far too close for comfort. After a terrific explosion and huge accompanying dust cloud, Harries moved forward to clear the compound, now destroyed. The only sign of life was a few AK-47 magazines.

After what seemed like an age, Snazle arrived at the helicopter landing site. Several of the extraction party now became heat casu-alties themselves and lay panting in the dust. The most serious of the group were extracted by the team on the Chinook which thankfully offloaded some bottled water onto the dusty, rugged ground. Red-faced and dripping with sweat, those who remained gulped down the precious liquid. Snazle, exhausted himself, now gathered up the remainder of the extraction party with as much water as they could carry, and returned to company HQ in the little bullet-pocked compound.

It was obvious to David that they would be unable to push on further that day The Afghans were just as low on water and ammunition and similarly exhausted. David, in consultation with the battlegroup commander, decided that they would have to realign themselves with B Company, now consolidating in Rahim Kalay. It was demoralising to leave the positions that the Queen's Company had spent all day fighting for but, given the overall picture, there was little else that could be done. 1st Kandak spent that evening back where they had started the day before. They had lost seven men as heat casualties and had fought all day long, capturing most of their objectives and seriously hurting the Taliban, thereby preventing the enemy from contributing to the fight against B Company. The CSM was very annoyed that they had been forced into a situation where resupply and casualty evacuation had been so difficult. It had been a very dangerous day and they had come close to being cut off. Control of the ANA in these situations was extremely difficult and David – along with the platoon and section commanders – had done exceptionally well to maintain the cohesion of his little mixed force. David was very proud of his men, they hadn't let him down when they were really up against it and they had again done all that was asked of them in very difficult circumstances.

The following day 1st Kandak was ordered to take up defensive positions in the area of Rahim Kalay to help protect the engineers who were preparing to start the construction of the new base. There was sporadic contact throughout the day but, undeterred, the Sappers kept working and by 1820 hours the sand-filled bastion walls were 90 per cent complete, although some had already been damaged by enemy fire.

Day 3 of the operation saw ISAF forces trying to keep a relatively low profile. A large shura had been arranged in order to speak to the elders in the district. It was important for them to understand what ISAF and their own government were trying to achieve in the area. This was also an opportunity for the elders to air their concerns. The shura was well attended and was deemed to have been a success.

Sporadic attacks continued throughout the day, then in the late afternoon the 1st Kandak positions came under heavy fire from the

Green Zone with tracer from the Taliban DShKs raking across the company HQ positions. In the evening a large reconnaissance force, comprising men from Amber 63 and a platoon of hand-picked ANA men, moved into the Green Zone. Although they moved cautiously into this enemy-occupied area, dogs in the compounds gave away their presence by their incessant barking. The patrol eventually reached its limit of exploitation without incident, but then stumbled across an enemy command post. A confused and desperate firefight ensued before the patrol withdrew 1.5km to safety. Supporting fire from mortars and artillery rained down on the enemy positions and intelligence later suggested that around seven Taliban had been killed in the engagement. Most estimates suggested that 60 or so enemy fighters had been killed since the beginning of the operation, but intelligence indicated that they had been reinforced by foreign fighters, which may well have explained the continuing and determined attacks.

With the new base finished, the clearance of Adin Zai was scheduled to begin on the morning of 12 July. The plan was for elements of 2 Mercian to cross into the Green Zone and to clear a series of compounds followed by 1st Kandak's advance from the west, and a simultaneous assault from the Czech group providing some support. As it transpired, the Czech group didn't cross the line of departure until much later. The Grenadiers and their Afghan colleagues commenced their part of the operation on time at 0600 hours. Suddenly, at 0620 hours while still on their line of departure, they were engaged in contact by the enemy and a fierce firefight ensued. Amber 61 suppressed the enemy position while Amber 63 moved to assault the well-prepared compound, which was eventually taken with the aid of close air support. Thick dust hung in the air and the sound of small arms fire was all around. Amber 61 discovered a booby trapped anti-tank mine in the compound and as a precaution everyone was evacuated.

It was now the turn of Amber 63 to assault the next batch of troublesome compounds and Harries led his troops, both British and Afghan, around to the right using whatever cover was available. It was clear that there were a number of enemy firing points and that they intended to defend their positions strongly. Harries planned a dash across open ground to a compound wall where there would be some protection from enemy fire. From here he would launch his assault. The young officer sprinted off followed by the remainder of the platoon. Sergeant Clint Gillies ran as fast as he could across the ploughed field, followed by Sergeant Paul Fear carrying the heavy GPMG. They were soon spotted by the Taliban who rained fire down on the sprinting soldiers as they headed for the protection of the clay compound walls. The crack of 7.62mm AK rounds was terrifying and the ground around them was flicked up by the impact of the bullets. A loud whoosh signalled the arrival of an RPG rocket which actually passed between Fear and Gillies; it was so close that the two NCOs could smell the burnt propellant from the projectile. Amazingly, everyone reached the cover of the wall safely. Fear and Gillies glanced at each other, wondering how they had survived. The two Queen's Company veterans had served together in various conflict zones for almost 20 years, but they had never encountered anything like this before.

Harries now launched his assault into the first compound and Fear pushed to the forward edge of the compound wall with his group of Afghans. From here he would be able to cut off the withdrawal of the enemy and could prevent any attempts at counter-attack. Glancing to his left he noticed a crouching figure in the ditch. At first he thought it must be a friendly but then realised that no one was forward of his position, and in a few seconds he brought his weapon to bear and shot the crouching enemy fighter. Among the sound of shooting and bursting grenades, punctuated by British voices shouting instructions Fear soon heard shouts of 'clear', signalling that the assault on this first compound had been a success. Harries led the platoon towards their next objective, but Amber 63 were soon under fire again. The air was filled with smoke and the smell of cordite as the

Grenadiers replied in kind. Harries used the opportunity to sprint across open ground to the next compound wall and once there he beckoned Fear and Lance Sergeant Roper to join him. Staff Sergeant Blow, the Royal Engineers NCO, threw smoke into the open ground and the party sprinted for their lives. A group of trees to the left was being shredded by enemy fire and Fear expected to be hit at any moment. Miraculously they once again reached the safety of the compound walls.

Harries led the way towards a mud wall and then into a small tree line, where he expected they would launch the assault onto the enemy position. A sudden burst of fire dropped Harries who let out a sharp yelp as he fell into a ditch. Fear and Roper crawled towards the young officer, when a second burst of fire churned up the undergrowth around them. Roper jumped up and dashed to Harries where he started to give first aid. The officer had been shot through the thigh but was conscious and coherent. Roper shouted for Fear to come and help and was dismayed to hear the reply, 'Hang on, I'm just bandaging my hand, I've been shot!' A bullet from the second burst had struck Fear's GPMG before ricocheting into his right hand.

At the company headquarters' location, David was concerned to hear Gillies scream down the radio, 'Sunray down!' and he feared the worst, but Harries himself was soon heard stating that he had been shot in the thigh and was applying a tourniquet. Gillies and Blow organised the extraction of the injured platoon commander using a lightweight stretcher. The two NCOs dragged Harries, using their belt buckles and under intense fire from several locations. It was exhausting work; they had to cover largely open ground for more than 350 metres, taking a good 30 minutes. Fear, having bandaged his damaged hand did his best to fire his GPMG left-handed until the group withdrew after Harries. Snazle was eventually able to load the injured men onto his quad bike and extract them via the MERT Chinook. Gillies and Blow drank some water and then returned to the still-raging battle.

At around the same time that Harries and Fear were wounded, nearby the Londons and Grenadiers attached to B Company of 2 Mercian were also assaulting a compound. During the firefight Guardsman Daryl Hickey was shot and severely wounded. Hickey was a Queen's Company man through and through. A popular and experienced soldier, the 27-year-old Birmingham man had been sent to bolster the Grenadiers working with the Londons and was supporting his comrades in the Queen's Company when he was hit. Hickey later died of his wounds.

Concurrently, the rest of the Queen's Company and 1st Kandak group had also been in constant contact. The enemy had been caught largely by surprise: in some of the compounds contained their half-eaten breakfasts and still-warm chai. The ANA showed great courage and efficiency as they fought through the seemingly endless compounds. Close air support was on station but the confused and close quarter nature of the fighting made dropping bombs a very risky business. On at least one occasion the British and Afghan troops narrowly escaped death when a bomb was dropped too close for comfort. A mud wall provided the cover necessary to survive the impact, but the shockwave was sufficient to flatten everyone and to squeeze the air out of their lungs. Guardsman Mcbride was thrown to the floor and was temporarily blinded by the force of the explosion. The same unfortunate man, having regained some vision, was then hit by the impact of another aerial bomb shortly afterwards. On this occasion he was on a compound roof when the shockwave forced a partial collapse of the building. Mcbride fell through the roof, incredibly without sustaining further injury. He emerged, covered in dust, dazed but none the worse for the experience with the exception of his damaged eyesight.

By 1430 hours a further 18 compounds on the northern edge of Adin Zai had been cleared by the OMLT and ANA troops and more enemy dead were discovered. Unfortunately, Captain Tim Badham had broken his arm in three places, further depleting the seriously undermanned Queen's Company OMLT group. David and Hennessey were now the only remaining officers in the company.

The fighting continued well into the evening and the troops remained in defensive positions overnight with all available surveillance assets used to monitor enemy activity. In spite of their heavy losses, the Taliban showed no sign of withdrawing. Intelligence suggested that they were continuing to reinforce and planned to press home further counter-attacks. British artillery and mortars continued to hammer the enemy whenever movement was detected and several air strikes were called when the Taliban were seen infiltrating. A limited resupply was organised under the cover of darkness and the CSM made sure that there was enough ammunition and water available for the following day. The exhausted troops slept when they could, their heads resting on their equipment, weapons close at hand.

Day 7 of the operation on 13 July started badly. An ANA soldier was attacked by a dog and one of his colleagues decided to solve the problem by shooting the dog. Unfortunately the bullet missed the animal but hit the soldier in the leg. It then ricocheted and hit the ANA platoon commander in the stomach.

1st Kandak continued its clearance of Adin Zai. To their relief no enemy resistance was encountered. By 1600 hours the Grenadiers were moving north to link up with the Czech special operations group, when they were ambushed from several different positions and an intense firefight ensued. Mortar fire was also received and the enemy base plate was located on the eastern side of the Helmand River. Close air support subsequently delivered nine 540lb bombs onto the enemy, to devastating effect. Adin Zai was now cleared of enemy fighters and a defensive line was established.

The following day a shura was held with local villagers. It was heartening to hear that there had been few civilian casualties and that the enemy were believed to have lost around 60 men, who had been killed on the previous night. Over the following days, farmers returned to their fields and families started to return to Adin Zai. Aggressive patrolling continued and there was plenty of enemy activity. The Taliban had been kicked out of the objective areas and the operation was deemed a success, but the fighters were already reinforcing in the Upper Gereshk valley. They had taken terrible losses over the previous week and intelligence suggested that a number of important commanders had also been killed. The exhausted 1st Kandak were eventually withdrawn on 16 July, more than three months after the Queen's Company group had left for Operation Silicon. When they eventually arrived back in Camp Shorabak, David's men were bearded and worn. Some of their uniforms were in rags and, without exception, they ruefully acknowledged that they stank. Their comrades back in Shorabak said that the smell was an indication that the Queen's Company were back. The truth was that everyone was relieved that Operation Tufaan was finally over. Unfortunately, the fighting was not yet over for 1st Kandak or the Queen's Company.

HELMAND RIVER VALLEY NORTH & OPERATIONS

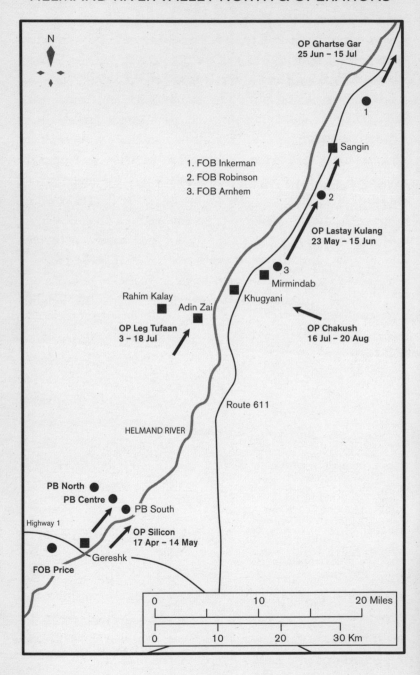

15

OPERATION CHAKUSH

As the fighting raged around Rahim Kalay and Adin Zai, 3 Company were relieved in Garmsir after three months of hard fighting. They flew in Chinooks back to the unfamiliar surroundings of Camp Bastion for an all too brief respite from the harsh desert environment. Task Force Helmand was already in the advanced planning stages of a further strike operation. A continuation of the same effort, which was to further force the Taliban away from Gereshk and to keep the pressure on them, Operation Chakush (meaning 'Hammer' in Dari) was to be led by Battlegroup South, which would move north for this effort. The purpose of the operation was to seize a crossing over the Nahr-e Saraj Canal and to punch into the Green Zone between Gereshk and Sangin. The thrust would come about 4 or 5km upstream from the area in which 1st Kandak and 2 Mercian had been fighting, but this time on the south side of the river valley. It was hoped that this would keep the enemy off balance and would force them to turn their effort away from the newly established base near Rahim Kalay. Battlegroup South would attack from the south and once the enemy were cleared from the objective area, another FOB would be built by the incredibly efficient Royal Engineers. More very tough resistance was expected, particularly in the main villages of Mirmandab and Hyderabad where intelligence suggested that around 150 enemy fighters massed.

Captain Rupert King-Evans found himself in command of 3 Company for this operation as major Will Mace had departed on his R&R. King-Evans received his orders from the commander of Battlegroup South. B Company of the Royal Welsh were to attack and seize the crossing before 3 company moved through them to further clear the area of Taliban. Second Lieutenant Howard Cordle was also present at the O Group. No. 2 Company OMLT were to lead an ANA manoeuvre element which would secure the bridge crossing site to allow B Company and 3 Company to push out and clear the area to the west and south-west. They were also to secure the lines of communication. It was felt that an ANA presence would be helpful in securing local consent and in intelligence gathering. 2 Company were also told to be prepared to assist the other assault elements in clearing the many compounds.

The attack was due to commence on the morning of 24 July and the men of 3 Company flew by Chinook into FOB Price from where the operation would be mounted. Most of the manoeuvre elements moved out by vehicle to an assembly area in the desert, to the south-west of the objective. 3 Company were a dismounted infantry company with no transport so on this occasion they were to fly direct from FOB Price to the assault area. Further orders and rehearsals were conducted in FOB Price and the troops did their best to relax before the flight. It was rumoured that the new FOB was to be called FOB Arnhem in honour of the brigade commander's roots in the Parachute Regiment. The officers selected DVDs to watch and before long they were settled down in front of *A Bridge too Far*. Some wag commented that they hoped FOB Arnhem wouldn't be a FOB too far.

Meanwhile 2 Company, led by Cordle, departed from Camp Shorabak together with their ANA troops who were loaded onto the backs of their Ford Rangers and five large trucks. They spent the night of 22 July in FOB Price with 3 Company and prepared for their move to the assembly area. Unfortunately for Cordle and his men, they were way down the order of march to the desert assembly area and it was early the next afternoon before the joint British and Afghan group left the FOB. The route to the assembly area would take them straight through the centre of Gereshk at the most dangerous time of the day for an attack. It had only been three weeks since the fatal suicide attack on the Queen's Company on this very route and intelligence suggested very strongly that the enemy were looking for an opportunity to repeat their success. Sergeant Hill and Lance Sergeants Bonsell, MacDonald and Shadrake were all still recovering from the serious injuries sustained in the ambush. Only Edgell and Mizon had been able to rejoin the company. The rest of the battlegroup had already passed through and any casual observer could hardly have missed the fact that something was going on.

From his seat in the WMIK, Cordle fired a series of mini-flares above the oncoming traffic. This signal was well-understood and the slow-moving Afghan cars and trucks duly pulled over to allow the British convoy space and some degree of security. Activity in the market ground to a halt and the convoy passed through at speed. The troops continued to fire the flares to warn any vehicles to keep away and this effort was largely successful, until a blue van was seen about 150 metres distant. Unlike the other vehicles it continued to head towards the British vehicles. The van was quickly spotted and more flares were fired but to no avail. The distance between the converging vehicles closed quickly. In Cordle's vehicle, Lance Corporal Archetinis drove with pistol in hand while Guardsman M manned the GPMG from a standing position behind. The blue van was now accelerating and its course was sure; the driver seemed intent on heading straight for the convoy. Warning shots cracked above the oncoming vehicle but still the single occupant maintained what appeared to be a determined course. In a last-ditch attempt to avoid lethal force, Cordle shouldered his rifle and fired a single round through the windscreen of the van on the passenger side. This had absolutely no effect and a burst was fired from the GPMG directly at the driver. The windscreen shattered and the van finally started to slow as the driver slumped dead over the steering wheel. Archetinis swerved to avoid a collision before pulling back onto the main road. It was too dangerous to stop and Cordle couldn't risk a secondary attack by going static in this urban area. He kept the convoy moving through Gereshk and by the time Sergeant Brooks passed the scene at the tail end of the convoy both the vehicle and the driver had been spirited away. The incident was reported immediately but no one was sure whether this had been a suicide attack or an attempt to ram a British vehicle by a deranged individual. Either way he wouldn't be doing it again.

2 Company eventually arrived in the desert assembly area without further incident. A huge leaguer had been formed in an area surrounded by high ground; the site had been well-chosen. Estonian troops provided an armoured guard throughout the night of 23 July, which allowed a fairly relaxed atmosphere inside the perimeter. The BRF were here too and the 2 Company men took the opportunity to catch up with their Grenadier pals, some of whom they had not seen in months. The BRF troops were bearded and looked worn. They had clearly been in the thick of it for some time.

Back at FOB Price two newly arrived colour sergeants joined 3 Company for the operation. Colour Sergeants Williams and Boak had just completed postings at the Royal Military Academy Sandhurst and this would be their first taste of action in Afghanistan. Williams joined B Company of the Royal Welsh as a liaison officer responsible for ensuring that 3 Company linked up smoothly, acting as a guide before rejoining the company. He moved off by vehicle into the desert with the Welsh infantrymen. CSM Matt Robinson organised a range so that the company could double-check the accuracy of their weapons and he managed to scrounge a quad bike and trailer from somewhere. This was an incredibly useful piece of equipment and would prove invaluable.

King-Evans delivered his final orders to the company on 23 July and next morning 3 Company were lifted from FOB Price in two Chinooks. It was a certainty that the enemy would see the aircraft descending in the half light, so the Grenadiers readied themselves for a sharp exit. It was vital that everyone was disembarked in seconds as the Chinook pilots would not place the aircraft at risk by sitting on the landing site. The rear ramp was open and the heavily laden infantrymen sprinted down the ramp in two files into the downwash of the rotors; the heat from the engine exhausts was almost unbearable and the smell of burnt aviation fuel filled their nostrils. Ahead was nothing but dust; they ran blindly into it so as to allow those following to clear the aircraft. As soon as the last running man stepped off the ramp, the aircraft rapidly lifted into the sky.

Each of the Grenadiers now lay in the dust with his weapon in the shoulder. They were vulnerable at this stage and were still disorientated. As the roar of the helicopter engines grew distant and the dust started to settle, the landscape slowly came into view. King-Evans checked his map and satellite navigation device in order to satisfy himself that they were in the right place. To the north, the Green Zone could be seen, a little less than 1km away. The distant sounds of battle could be heard and over the top of the lush vegetation black smoke and brown dust rose. The Royal Welsh attack had commenced on time and it looked as though they had a fight on their hands already. 3 Company were not yet complete; the Chinooks had to return to FOB Price to pick up the rest of the company and King Evans used the time to orientate himself and to ensure that the company fanned out to cover any likely threat. After about half an hour the familiar beat of the Chinook rotor blades was heard and it wasn't long before the huge green machines were once again showering everyone in dust. After only a minute or so they once again vanished into the sky leaving a completed 3 Company lying in the desert. Looking around the dispersed company, King-Evans saw men adjusting their equipment and pulling packs onto their backs. One or two felt the need to relieve themselves before moving off. It was unclear if this was due to apprehension, but their comrades ribbed them about it anyway. The fighting in the distance looked to have intensified and more black smoke was rising above the trees.

The plan involved the Grenadiers being ferried forward in a number of Viking armoured personnel carriers and before long the tracked, box-like vehicles arrived from the direction of the battle. The little armoured carriers were crewed by the Royal Marines Armoured Support Group and the men of 3 Company clambered aboard, grateful that they would not have to hump their heavy equipment across the desert. Soon, King-Evans and his men found themselves at a group of compounds on the edge of the Green Zone that would later become known as 'Zulu Crossing'. The company headed across a hastily constructed bridge and as they did so they encountered two medics carrying the body of a Taliban fighter in the opposite direction. It was an unwelcome reminder of what lay ahead. King-Evans's plan involved a rapid forward passage of lines through the Royal Welsh, but the picture on the ground was a confusing one, making a guide absolutely vital.

Right on cue Williams appeared out of the dust and reported to King-Evans. He led 3 Company through the tired-looking Welsh soldiers and they then fanned out. Captain Stuart Jubb led with his platoon. The threat from mines and IEDs was very high so Jubb's men were themselves led by a section of Royal Engineers with mine detection equipment. Progress was understandably slow and King-Evans became frustrated; he was keen to move forward before the enemy reorganised themselves. He talked to Jubb on the radio and the two agreed that they would press forward without the engineers. The company headed for the village of Murmandab, which they were to clear. The platoon commanders ordered the men to fix bayonets in anticipation of a tough close quarter battle. It was clear that the inhabitants had fled the village when they saw the British troops coming. Spilt food and upturned cooking pots lay around together with other hastily discarded items. Animals were still tethered around the compounds as a further indication of the rapid evacuation of the village. It was still early morning but already nearly 40°C as the troops cautiously advanced through the maze of compounds, some of which had been hastily padlocked. There were sudden bursts of automatic fire on the left of the company as the fire support group under Colour Sergeant O'Halloran came under attack. Then a brief pause before the sound of outgoing fire joined the cacophony of gunshots. A familiar pattern of cat and mouse ensued as the Taliban tried to halt the British advance through the village. Progress was slow as each compound was cleared and the enemy did their best to inflict casualties before being forced back. King-Evans had sensed that the Taliban resistance, although determined, was not well-organised and he thought they had been caught off balance by the unexpected British assault.

Then as the late afternoon approached, resistance suddenly increased. 3 Company came under intense small arms and RPG fire securing a large compound that King-Evans had nicknamed 'Forecourt'. It seemed as though the Taliban had managed to reorganise after the initial shock and now started to arrive at an alarming rate. The familiar whoosh of the rockets was unnerving but most of them fortunately flew past overhead, through the trees, to explode harmlessly against the compounds to the rear of the company. At one point there were more than 20 explosions from these lethal rockets in a 15-minute period. 3 Company were pinned down from the front and were unable to advance. One of the RPG rockets burst through a compound wall and embedded itself at the feet of two startled Guardsmen without exploding. It was an incredibly lucky escape. The company did, however, sustain its first casualty of the operation when Williams was struck in the hand by shrapnel from a bursting RPG rocket. The wound was serious enough for the NCO to be evacuated on only his first day in action. Williams was devastated to leave the company but there was no choice, his injuries would limit his participation in the operation and he knew the medics were right.

King-Evans crouched down in the protection of a compound and did his best to direct the battle that was raging around him. Lance Sergeant Poxton, his signaller, relayed information to and from the dispersed platoons, each of which was fighting its own battle. The attached FST called in artillery and air support. Brown controlled the buzzing Apaches that swept in from above. Despite the awesome firepower available to ISAF, the Taliban continued to fight fanatically for every piece of ground. To the left of the company HQ location, a tree was slowly being shredded by the weight of enemy fire that was sweeping across the top of the compound. Poxton and his commander watched as two RPG rockets sped through the trees without encountering anything hard enough to detonate the warheads.

Back at the bridgehead, 2 Company and their Afghan charges had moved forward and relieved the Royal Welsh. They now set about securing the area and protecting the lines of communication. Abandoned enemy positions could be seen: trenches and rat runs linking fire positions surrounded the bridge. Dead Taliban lay on the ground, testament to the hard fighting that had already taken place there. Cordle decided to occupy a nearby compound and the ANA got to work clearing the various buildings and rooms. Almost immediately the sharp-eyed Afghans discovered IED components and explosives. At the bridgehead they expertly discovered a booby trap which they removed without help from the ngineers, taking the live device straight to the horrified Cordle, who rapidly ordered its removal.

Cordle sent two of his Afghan platoons, together with their UK mentors, over the crossing to the north side where they established defensive positions. He kept his company HQ on the south side in the compound they had occupied, close to the crossing point. The OMLT vehicles with their mounted heavy weapons were positioned where they could support the platoons on the far side of the bridging point. The bridge was vital as resupply would be conducted using this point and any casualties would come back this way. Cordle was satisfied that the crossing was as secure as he could make it. In the afternoon, the Taliban, recognising the importance of the crossing, launched some tentative attacks using both mortars and small arms fire. Two enemy fighters were shot and killed as they attempted to skirmish towards the northern platoons and the vehicle-mounted heavy weapons brought effective fire to bear.

In the late afternoon, B Company leapfrogged through 3 Company on the northern side of the crossing and continued the fight in Murmandab. The Grenadiers formed a secure perimeter and settled down for the night. Those who were able caught some sleep where they could. They all knew that the following days would be just as challenging as today. King-Evans received his orders over the radio from the battlegroup commander and, as he scribbled into his little notebook, he was conscious of the bursting illumination rounds fired by the mortars. Their ghostly flickering lit up the area and volleys of sporadic machine gun fire were heard throughout the night.

At dawn the Grenadiers pushed up and again passed through B Company of the Royal Welsh. They pushed through the village, clearing compounds as they went. Above them the deadly Apaches were ever present and fired Hellfire missiles and bursts of 30mm cannon at the enemy wherever they were seen. The Taliban were being steadily forced back and B Company could be heard fighting at stages too. As the fighting around Murmandab developed, the BRF became engaged in a fierce fight a little further to the west and closer to the familiar village of Kughanyi, which was situated just to the south of the main Helmand River. The reconnaissance troops had been in contact with the enemy since the outset. They had secured the forming-up point and had had little rest since. At first light they received the news that they were to move into the dreaded village where they had seen such tough fighting only weeks earlier. Their approach had initially been quiet with no sign of the enemy, but this suddenly changed as the BRF got closer. A massive volume of automatic fire rained down on the British troops and volleys of RPGs exploded around their vehicles and dismounted positions. Almost immediately CSM Ian Farrell heard the shout 'Man down!' There were two casualties and he once again ran through a hail of gunfire to get to the wounded men. Sergeant Barrow had received fragmentation wounds to his left leg and arm. The NCO was unfortunate enough to also have been shot in his left shoulder. Lance Corporal Donovan too had received fragmentation wounds and would need rapid evacuation.

As Farrell reached the edge of the village the BRF received more casualties. This time the FST was hit and there were three more casualties to extract. The FST was a vital part of the BRF who were now unable to communicate with the helicopters. Somehow Farrell was able to extract all five casualties, some of whom had quite serious injuries, and the fighting continued. Kughanyi was a name that the BRF wouldn't forget any time soon.

The battlegroup commander was concerned that the northern flank of the operation was vulnerable. With 3 Company and B Company of the Royal Welsh pushing steadily to the west and south-west, he saw that there was an opportunity for the enemy to regroup to threaten the bridgehead and the area in which the engineers were to build the new FOB. He ordered an ANA offensive into this northern area. The OMLT and ANA were thinly spread, but Cordle quickly reorganised his small force. The company HQ together with an Afghan platoon was left to secure the crossing and Cordle moved north with two light platoons and the bulk of the 2 Company OMLT. He was disappointed that the Afghan commander chose to remain behind at the bridgehead. The advancing troops soon found themselves moving through abandoned enemy positions and the ANA were quick to collect the weapons and ammunition found in the various trenches. Although it was desperately hot and the going was necessarily cautious, the 2nd Kandak men made good progress.

Task Force Helmand then decided that another manoeuvre group of ANA soldiers was needed to exploit the gains that had so far been made. Back in Shorabak there were no troops available other than the recently returned men of 1st Kandak and the Queen's Company who were now preparing to replace the Ribs in Sangin. The OMLT group was ordered to find a company of ANA men that would be pushed forward to reinforce the hard pressed 2nd Kandak troops. It was another unpleasant surprise for the Queen's Company OMLT men who were still recovering from their fight in Adin Zai a little over a week earlier. Given their recent casualties and leave that had been taken between operations, there were no officers except Captain Rob Worthington available. Worthington gathered up ten OMLT mentors and 60 ANA soldiers and made preparations to join Operation Chakush.

For those already involved in the fighting, the night of 25 July was again spent in the compounds of Murmandab and the surrounding area. At 0530 hours on 26 July, 3 Company began their advance to contact to the west. King-Evans had barely stepped out of the compound he had occupied overnight when the lead platoon came under contact. The initial sporadic fire quickly intensified until a heavy weight of sustained and accurate fire was being rained down upon the company. Even company HQ was receiving fire with rounds churning up the ground, shredding the undergrowth nearby and cracking overhead. Lieutenant Nigel Torp-Petersen and O'Halloran sent situation reports by radio to King-Evans, giving the locations of identified enemy positions to their front, which were as close as 100 metres in some cases. Javelin missiles were being fired at the more distant positions but even these were very close. The FST called for air support and the remainder of the company returned a heavy rate of fire onto the enemy. The sounds of battle were deafening as incoming rounds cracked overhead like whips and outgoing fire from the British weapons added to the noise. The situation was both alarming and confusing. No one was able to move very far because the enemy fire was so effective. More Javelins were fired and the missiles were seen heading skywards, before shooting to the ground at great speed and exploding in thick brown clouds of dust.

King-Evans suddenly heard alarmed shouts of, 'Medic!' The urgency in those voices told him that this was serious. Shortly afterwards O'Halloran's voice boomed across the radio to inform company HQ that his group had sustained a 'T1' casualty. This meant a very serious and life-threatening injury. The NCO followed up this information with the news that the casualty had a gunshot wound to the right shoulder area. Robinson heard this news at the same time and, together with his little group and the company medic, he ran to the location of the casualty. King-Evans was already requesting the MERT and an immediate emergency casualty evacuation. After what seemed like an age, but in reality was a matter of a few minutes, the CSM's party was seen heading back towards their secure compound. They were carrying an unconscious casualty and King-Evans was able to recognise the soldier as Guardsman David Atherton. He followed them back into the compound where he informed the battlegroup headquarters of the situation. The young soldier needed to be evacuated quickly and he was loaded into the trailer of the CSM's quad bike. Robinson raced off towards the crossing where a doctor was waiting.

The battle was still raging and King-Evans directed his platoons in their continued attempts to capture the strongly held compounds to their front. He gave detailed directions to the helicopters overhead which fired Hellfire missiles into the occupied buildings to devastating effect. The situation remained very confusing and King-Evans decided that he needed a face-to-face briefing from his 2 Platoon commander, Torp-Peterson. After getting some covering fire from the FST, he sprinted off down a narrow irrigation ditch. After a few moments he was able to locate the young platoon commander who had moved to the rear of his in-contact platoon to meet him. Torp-Peterson was in good heart despite the heavy fighting that was going on all around. He pointed out the detailed positions of his sections and the identified enemy locations, on his map. As the two officers knelt in a ploughed field and discussed the tactical situation they suddenly came under fire. An enemy fighter had seen them and now loosed off a whole magazine. Bullets kicked up the earth and passed between the two men. Neither was hit but they dived for cover, conscious that the next burst of fire could end their lives. After returning fire, the officers separated once again, King-Evans returning to his company HQ. It was now critical for the deadlock to be broken and Torp-Peterson pushed forward a section under Lance Sergeant Bayliss to break into a series of difficult compounds known as Lima 1-5. Bayliss was successful, although it was necessary to prevent the enemy from re-entering the compound which was ablaze. The fire was so serious that he was forced to send one man to try and fight it. So intense had been the fighting that one of Bayliss's men was forced to strip and clean his light machine gun because it had jammed due to the sheer amount of ammunition fired through it. Once Lima 1-5 was secure, King-Evans moved his company headquarters into the compound with 2 Platoon. Robinson had now returned after dropping off the wounded Atherton and followed with the essential water and ammunition in his quad trailer. At this point Poxton, who had been monitoring the battlegroup radio net, called King-Evans aside. The battlegroup medical officer had confirmed that Guardsman Atherton was dead and Poxton broke the news to his company commander. King-Evans's worst fear was confirmed. It was terrible news and the young officer's heart sank. He decided not to share the tragic information just yet.

The surrounding area had now been secured and the company moved into a series of more defensible compounds that forded better views of the objective areas. The company had been in contact for 12 hours in the intense heat and were now forced to refill water bottles from the wells in the compound. The liquid was milky in colour with insects and other foreign bodies floating in it. The troops added purification tablets and hoped for the best. As they settled into a night routine, King-Evans briefed his officers and seniors for the next period. He broke the news of Atherton's death. A heavy atmosphere settled over the company position that night. Atherton, or 'Jaffa' as he was known because of his red hair, was a popular character and would be sorely missed. A sad and thirsty night was spent in the glow of illumination flares.

July 26 had also been a busy day for 2 Company. Cordle had again pushed forward through some thick maize fields and once more encountered the enemy as soon as he approached the next series of compounds. On this occasion 107mm rockets were fired directly at them; the low trajectory meant that the rockets hit the ground and slid along, straight past the Grenadiers, ploughing furrows as they went.. The ANA troops, led by Cordle, managed to get within 50 metres of the enemy position where the voices of the Taliban within could be clearly heard. Cordle decided to withdraw and use air power. Two Apaches destroyed the enemy and 2 Company, together with the 2nd Kandak, were able to move in and clear the area. Monitoring captured radios, the troops listened to unanswered calls from more distant Taliban asking where the 'infidels' were. The OMLT and ANA now consolidated for the night just as their colleagues in 3 Company were doing to the south-west.

On 27 July, King-Evans led a strong fighting patrol through a compound flattened the previous day by two 500lb bombs called in after close surveillance had indentified a Taliban position. There wasn't much left but, to his amazement, an old man appeared. For a moment King-Evans feared that some civilians may have died, but the old man put his mind at rest. He told the British soldiers that no fewer than nine Taliban fighters had been killed in the air strike. He also pointed out the blood-stained ground, where he said a further six fighters had been killed by the Apache helicopters the previous day. A more detailed search of the surviving buildings revealed a cache of munitions, including RPG rockets, explosives and IED triggers. Bizarrely there was a British miniature UAV hanging on the wall like a kind of trophy. The troops returned to their compound and enjoyed a relatively peaceful night.

On 27 July the small Queen's Company group headed by Captain Worthington arrived to reinforce the exhausted ANA group with Cordle. After a briefing from the battlegroup commander at his HQ, Worthington led his men off towards the bridgehead where he met Lance Sergeant Alexander who had been organising the logistic effort for Cordle's group and was able to update him on recent events and to warn him that the fighting here had been intense for some days. The small group of Queen's Company men and their 60 soldiers from 1st Kandak moved off to find their 2 Company colleagues. At around 1100 hours Worthington located a guide positioned by Cordle and the two groups linked up. When Worthington and his men entered the compound held by 2nd Kandak and their mentors, Cordle was delighted to see them; it had been a tough couple of days and the sight of more Grenadiers and some fresh ANA troops was very welcome.

At around 1400 hours, Cordle led his group out of the compound. Worthington and the Queen's Company men were eventually to follow, but for now they provided overwatch for their colleagues. The plan was to push out and link up with the Royal Welsh who were now about a kilometre to the south-west. The ANA troops, together with their OMLT mentors, moved in extended file across some low wheatfields towards a dense belt of vegetation and the compounds beyond. There was no sign of activity from the front and Cordle was only about 50 metres from the vegetation belt when all hell broke loose. The Taliban sprang their ambush, unleashing a hail of automatic fire from multiple firing points. RPG rockets impacted, throwing up clouds of dust, and fire rained in from the north, north-west and west. The ANA troops, caught in the open, were vulnerable from this close range attack and four were hit immediately. The soldier just ahead of Cordle was shot through his helmet and slumped forward, dead. Cordle hit the ground hard and returned fire to his front. It was clear that some of the enemy were as close as 30 metres away; to try and rush them or to run would invite certain death.

The Grenadier officer crawled into the relative safety of a nearby irrigation ditch from where he engaged the enemy who were now moving forward from the compounds. The ANA troops had scuttled into any available cover and most had managed to withdraw to the safety of the compounds to the rear. Unfortunately, none of them had thought to advise Cordle of their planned actions and the young officer was now left isolated and in close contact with the enemy. His position was under intense fire which he returned desperately at the assaulting Taliban to keep them at bay. The Royal Artillery FST were also still in the fight, as was Archetinis who fought his way forward to join his platoon commander in the ditch. Behind them in the compound,

Worthington organised supporting fire and tried to reorganise the bulk of the ANA troops who had withdrawn with their casualties in something of a panic. It was clear that they would not go back out, they were tired from several days of hard fighting and the ambush had badly shaken them.

Brooks did a sterling job coordinating fire support to the north, as Cordle and Archetinis fought for their lives. Ammunition was running low and the situation for the two men was becoming desperate; they feared being overrun by the enemy. Bombardier Greenland from the FST crawled forward with an ammunition resupply and joined the two Grenadiers in the irrigation ditch, directing artillery and air strikes onto the enemy, who could be seen on the screens of the FST's hand held laptop which relayed live footage from a UAV above. For several hours Archetinis and Cordle fired more than 20 magazines of ammunition between them and the closest of the air strikes saw a 1,000lb bomb dropped into the compound closest to the group no further than 50 metres away. The skill of the FST and their ability to direct such accurate strikes from the air was undoubtedly the reason why the little group was still alive. Greenland in particular had shown great courage in getting himself into the most forward of the threatened positions.*

Meanwhile, Worthington had been directing heavy supporting fire from the roof of his compound which had also been under assault for most of the afternoon. He had requested armoured support in order to extract the beleaguered little group from their precarious position. After some difficulty, a couple of Viking carriers arrived. Worthington directed them towards Cordle's position and they pushed forward. Their presence, together with a couple of Mastiffs on the other side of the canal, was enough to worry the enemy and Cordle's group were able to withdraw into the safety of

---

* Greenland was subsequently awarded the Military Cross for his actions.

the compound they had left some hours earlier. The ANA casualties, some of whom were very serious, were loaded into the Vikings and were extracted back to the crossing point. The FST, now unrestrained by the presence of coalition troops to the front, unleashed a stream of air and artillery strikes. These had a devastating effect on the enemy who were soon destroyed. It was estimated that more than 30 Taliban had been killed in the area.

The next day 3 Company was ordered to move south in order to relieve a platoon from another NATO nation that was guarding a bridge at a crossing point known as 'Mike 119'. The platoon had taken over from the BRF who had initially occupied the compounds in the area and had taken some serious fire from the Taliban positions nearby. The BRF had been genuinely surprised at the elaborate defensive positions they found at M119. There were deep communication trenches, bunkers and tunnels under the road. Firing slits had been carefully prepared and the positions were well camouflaged. Fortunately the Taliban had been driven off and the coalition soldiers were now in control of the area. King-Evans was very aware of the potential for friendly fire casualties in this type of operation. He would be approaching a position occupied by NATO troops in an area that was still being contested by the enemy. He spent a great deal of time on the radio making arrangements with the NATO troops to ensure that the two forces didn't engage each other.

The handover of the crossing point was conducted swiftly and the NATO force departed. Within a few minutes an alert Guardsman spotted a suspicious wire and raised the alarm; he was convinced that it was a command wire and the ammunition technical officer was tasked to deal with the suspect device. He confirmed that a huge IED had been planted under the bridge and he was able to make it safe. The NATO soldiers had guarded the bridge for three full days without noticing the bomb planted beneath it. Jubb was left in command of the crossing site and King-

Evans led the rest of the patrol back towards their original compound. As the patrol moved away from Jubb's newly occupied position and across the open ground, they suddenly came under fire from the rear. The crossing point, too, was under very heavy fire. The patrol ran as fast as they could for the cover of the nearest compound. The air was thick with bullets and they seemed to be whipping about all over the place. This was the heaviest fire most of the Grenadiers had been under so far and they were amazed that they made it to cover safely without taking any casualties.

Worthington and the Queen's Company OMLT group was able to link up with the Royal Welsh on 28 July. Cordle and his men remained in situ to reorganise. The Welsh troops had been fighting every bit as hard as the Grenadiers and fighting was still taking place when Cordle and his troops joined them. The Welsh had their vehicles and heavy weapons to hand which gave them a huge advantage. In the following days the Queen's and 2 Company OMLT groups were able to work alongside each other patrolling and continuing to push back the Taliban, who, despite huge casualties, continued to push and oppose the occupation of this area.

Much skirmishing took place in the following days and US Apache helicopters frequently provided low-level support to the British and Afghan troops when they ran into trouble. On one occasion, the 2 Company men startled a section of Taliban who were digging defensive positions along the edge of a field. Surprisingly, the Taliban group reacted by throwing up their hands and immediately surrendering. As Cordle and his men tried to approach to take their prisoners, they came under heavy fire from enemy depth positions. The original group of Taliban now changed their minds, quickly snatched up their weapons and engaged the British and Afghan troops. A US Apache rapidly dealt with both groups by strafing the edges of the field. In another incident, Brooks distinguished himself by moving forward with light scales of equipment and personally putting two grenades into an enemy position which had been troubling the advance of the patrol. On the morning of 31 July, Worthington led the 1st Kandak troops out. Within 100 metres Sergeant Clint Gillies reported what he believed to be an IED. The device was eventually confirmed as a huge bomb which would have caused serious casualties, another lucky escape.

In a separate incident on 1 August, the advancing 2 Company men were again surprised at close range by Taliban fighters in a compound to their front and then by other positions to the flanks. A firing line was established and air support from the welcome US Apaches was again on the scene quickly. The first strafing run conducted by one of a pair of helicopters was highly effective, but the second proved to be disastrous. The US pilot got it badly wrong and engaged the British and Afghan troops with his 30mm cannon by mistake. The deadly cannon shells cut through the Afghans in the firing line, causing havoc. Cordle ran to assess the situation and as he did so he was himself caught in a burst of fire from the US helicopter. A round struck the ground, ricocheted and struck the young officer in the helmet, knocking him to the floor, concussed. His medic, Guardsman Lyne-Pirkis was not as lucky and received shrapnel injuries to the hand and left leg. Four Afghan soldiers were also hurt by the high velocity rounds and now lay on the floor bleeding from their injuries. Cordle gathered his senses and informed Brooks of the situation by radio before getting to work with the others to provide first aid to the wounded. The enemy were still in close proximity and attempted to exploit the situation by moving closer to attack the party trying to save the lives of the wounded. Greenland once again distinguished himself by moving up river and personally accounting for at least one enemy fighter moving in his direction. The casualty evacuation was a lengthy and difficult process but all five men were eventually extracted by Chinook from a desert helicopter landing site. Despite some quite serious wounds, all five of the casualties survived the experience.

These sporadic attacks by an increasingly disjointed and disheartened enemy continued for the following week. 3 Company conducted patrols around the area and were frequently under attack. Eventually they were told to move back from this area of the Green Zone to the newly completed FOB that had been constructed on a hill overlooking the area. They had done an excellent job and King-Evans was justifiably proud of his men. Before they left the Green Zone, each man put on a fresh shirt bearing the blue–red–blue flash of the Household Division. They marched out into the desert, heads held high in the knowledge that they had done well in the preceding two weeks.

Morale was high but this didn't last long. On arrival at the newly named FOB Arnhem, it quickly became apparent that this was a very sparse location. It consisted of a ring of bastion walls and a few internal walls which formed bays to be used as shelters. There was little else and the only respite from the burning sun was in the shade provided by the men's own ponchos. Worse still, the base had been built on a forward slope which made it vulnerable to incoming fire. Most of the other elements that had participated in Operation Chakush were now withdrawing to Camp Bastion or back to their permanent FOBs. Feeling a little abandoned in this most basic of outposts, 3 Compnay were unsure of how long they would once again be asked to hold the line. RSM Andrew Keeley was able to visit the exhausted men during this period. He could see how tired they were and knew that morale had taken a dip after the death of Guardsman Atherton. He was, however, surprised at how stoical the young soldiers were. Keeley talked to them about the days ahead and how they might be expected to repeat the actions of the previous weeks. He was very proud to hear them say, 'We know that, sir, bring it on.'

16

THE DEFENCE OF ZULU CROSSING

On 2 August, Captain James Shaw and a fresh group of ANA troops arrived at FOB Arnhem. They had driven from FOB Price to relieve Second Lieutenant Howard Cordle and his men and had been in the newly established outpost with 3 Company for only a few minutes when the Taliban subjected it to one of their increasingly frequent mortar attacks. Shaw was left in no doubt that the enemy in this area were not yet beaten and was surprised that the FOB had been sited on a forward slope, allowing the enemy to direct fire straight into the middle of the little base. After a short period he drove down to Zulu Crossing where he met with Cordle and Sergeant Brooks. Both men looked drained and it was obvious that they had seen some very hard fighting. A thorough handover was conducted before the outgoing OMLT men drove off, followed by their equally exhausted ANA troops.

The engineers had replaced the older wooden bridge with a sturdier steel version capable of taking the weight of an armoured vehicle. The defence of this important crossing was the responsibility of the OMLT and they now occupied three small compounds. Two were on the eastern side of the canal, defended by Grenadiers and ANA soldiers; the other, on the west side, was occupied by ANA troops alone. The compounds were about 100 metres apart and the crossing was just inside the Green Zone, overlooked by FOB Arnhem. Shaw and his callsign Amber 21 consisted of only seven men. They were accompanied by 53 ANA troops who were now split between the three compounds. The crossing point was obviously important to the enemy; whoever controlled it had the entry point into the Green Zone. Enough troops would be needed on the crossing to repel any determined attacks by the Taliban and this would restrict their patrolling activity. The prospect of holding a static position against Taliban mortar and rocket attacks did not appeal to anyone. There were no sand-filled bastion walls, just a few sandbags positioned on the roofs at key points overlooking the canal and the Green Zone beyond.

For the next couple of days, Shaw and his men concentrated on consolidating their defences and on learning the ground. It was important to liaise closely with 3 Company and he was relieved that he knew he could rely on the Grenadiers in FOB Arnhem for support if this were needed. On 6 August, Shaw pushed his first patrol out towards the enemy. The terrain was difficult and visibility was down to little more than 50 metres in places due to the thick vegetation. It was a tense patrol; several compounds were cleared but there was no sign of the enemy who seemed to be keeping their heads down. The next day saw enemy mortar attacks on Zulu Crossing but no real damage was done; it was nonetheless a tense time for those sheltering inside the mud walls.

Then, on the 9th things got much worse as mortar rounds and rockets were directed at FOB Arnhem and the crossing at intervals all day long. For the troops stationed at the crossing it was an unnerving time. The sharp crump and impact of mortar rounds could be heard about 100 metres away before more were heard just beyond. The Grenadiers wondered if they were being bracketed and if the next round would land on top of them, only to find the following explosions moving away from them up the hill beyond to the FOB and 3 Company. Those who have been under mortar fire will tell you that the feeling of helplessness in these circumstances is terrifying. For now, all that could be done was to try and locate the firing positions and to return the fire with British mortars. The Taliban were well aware of the consequences of being located so they frequently stopped firing and disappeared for a while, before resuming their attacks.

The following day was no better, with rockets being fired from the far side of the main river at a range of about 3km. The gunners in FOB Arnhem tried hard to locate the firing points to mount counter battery fire, but the Taliban moved after every salvo. Zulu Crossing seemed to be taking most of the fire on this occasion with rockets exploding less than 100 metres from the Grenadier and ANA positions. The Afghan troops had had enough and were determined to fire back. They produced their own mortar and prepared to fire blind into the distance. Shaw remonstrated with the Afghan officers and through his interpreter reminded them that they had no idea where the enemy was and that villages full of civilians were in the line of fire. An intense and farcical argument ensued with rockets exploding around the crossing all the time. In the end Shaw was forced to prevent the Afghans from firing by placing his hands over the barrel of the mortar and physically stopping the bombs being dropped into the steel tube. The Afghans eventually relented, but it was clear that they were unhappy. The constant indirect fire attacks were taking their toll on morale and that night there were disturbances among the Afghan troops. It seemed that there was some kind of internal dispute with allegations being made against the Afghan officer in charge. Things became heated and the OMLT men were powerless to intervene. No one knew what the outcome would be and the Grenadiers kept their personal weapons close that night. By morning the dispute had been settled and things were back to what passed for normal in the ANA. Nerves were frayed and everyone was dog-tired because of the need to keep a strong detachment alert at all times.

On the 12th, Shaw and his OMLT detachment joined 3 Company on a three-day operation to clear an area believed to be occupied by the enemy and known as 'Objective Waterloo'. Day one went off without any serious contact with the enemy, but was as usual tense and exhausting. The following day found the Grenadiers in contact with the Taliban; there was some accurate small arms and RPG fire taken before British Harrier jets destroyed the enemy. At one point during this contact Shaw was taking cover and trying to direct his sector of the battle while Sergeant Lance Owen did great work with the 51mm mortar. Shaw looked up and saw the ANA section commander crawling towards him. The Afghans had been difficult all day and he was pleased when the Afghan NCO slithered up beside him. Shaw expected the Afghan to ask for orders, but instead the wiry NCO simply asked if he could fire Shaw's SA80 rifle. This childlike conduct in the middle of a battle was enough to drive the British officer to distraction.

Everyone was tired and the relationship with the Afghans had become quite strained. At one point Shaw contacted Will Mace, now back in command of 3 Company, to say he thought he may have to withdraw the Afghans from the operation. The mission, however, was successfully completed before Mace withdrew the troops back into their FOBs. The OMLT men were relieved to get back to Zulu Crossing on 14 August, although there was little rest to be had with only seven Grenadiers to provide their own force protection.

The men of 3 Company had settled into a well-practised routine in FOB Arnhem. It was much like the situation they had become accustomed to during their difficult months in Garmsir. There were constant enemy mortar and rocket attacks during the day and offensive patrols to be carried out during both night and day. The patrols were designed to keep the enemy away from the crossing points and the settlements in the area. On the 16 August, Lieutenant Rupert Stevens, who had recently joined 3 Company from the 2 Company OMLT, was tasked to take out a patrol just before dusk which was designed to clear a number of compounds on the south side of the canal. Once this was complete he was to move on towards the forward line of enemy troops and locate a crossing point over the canal. Mace planned a larger company strike operation in the future and needed a suitable site at which an infantry bridge could be laid. Knowing that the dismounted patrol would be very vulnerable if it came under fire, Shaw offered to cover their flank using his OMLT WMIKs and heavy weapons.

Stevens led his patrol off as the afternoon sun started to fall below the horizon. They quickly reached their objective and cleared the compounds on the south side of the canal, which was about 5 metres wide. Having positioned his patrol with good arcs of fire, Stevens moved forwards with the engineer reconnaissance party and a section of Grenadiers commanded by Lance Corporal Richards. About 1km away Shaw's OMLT men watched through binoculars as Stevens and his troops moved forwards cautiously. The 3 Company men were suddenly ambushed from positions across the canal; the enemy was about 100 metres away and the Grenadiers could clearly see Taliban heads popping up above a small rise to take pot shots before dipping down again. There were at least three separate firing points and the enemy fire was accurate and intense. Guardsman Rowlatt immediately blazed away in response with his light machine gun as the exposed soldiers moved themselves into any available cover. Stevens realised that there was no real chance of assaulting the enemy positions as he would have to negotiate the canal and would be left exposed on the wrong side of the obstacle. He decided that the best thing to do was to withdraw to the compounds where he had left the rest of the patrol.

Meanwhile, the OMLT men opened up from their WMIKs. They had located the Taliban positions using the back blast of an RPG as it was fired. An awesome barrage of fire from two pintle-mounted GPMGs and a GMG tore into the enemy positions over the canal. Stevens took this opportunity to get his men to peel off and start moving back. As he did so, he noted that the enemy fire was rapidly chewing holes in the sparse cover available to him. At one point he saw Richards leap into the canal and wade chest-deep in water. He was amazed to see that on the other side of the canal the Taliban were also peeling off and starting to withdraw. He could clearly see four men moving in turn between pieces of cover. He noted that the four soon became two but he wasn't sure if this was because they had been hit.

Shaw and his men were enjoying their part in the battle as they rained fire down upon the Taliban who in turn responded with RPG fire. Fortunately, the OMLT men were out of range and they watched as the rockets exploded in the air about 200 metres short of their vehicles. Stevens was able to withdraw successfully with no casualties. He returned to FOB Arnhem with confirmation that the site could be bridged but that the enemy were dominating the opposite bank.

20 August started badly at Zulu Crossing and got worse as the day progressed. At first light the Grenadiers were startled by a bang and the sound of an RPG rocket whooshing over their compound. Everyone stood to in anticipation of a Taliban assault. When this failed to appear, there was a period of confusion. The ANA initially claimed that one of their sangars in Compound 3 over the canal had been hit by an RPG. After a while it became clear that there had been an accident with one of the ANA RPGs. The resulting explosion and back blast had caused four serious casualties and one walking wounded. Other soldiers who had been eating their breakfast had been showered with shrapnel. It was necessary to evacuate the casualties as soon as possible and this was done using the now all-too-familiar evacuation procedure. Shaw's already small detachment had now been reduced to 48.

Despite this tragic setback and terrible start to the day, Shaw decided to proceed with the planned activities. The regrettable accident served to remind everyone just how badly trained some of the ANA soldiers were and how the OMLT mentors needed to work constantly to improve their discipline and soldiering skills. The ANA had taken delivery of some SPG 9 recoilless rifles and it was clear that if they were to be used safely, some training was required. Shaw had organised a range day and planned to train the young Afghan soldiers on this new weapon as well as their RPGs and other small arms. About 500 metres south-east of Zulu Crossing there was a large, sandy feature which dominated the Green Zone and the surrounding area. It was known simply as 'Sand Hill'. To the rear of this feature was a large expanse of open desert. Shaw assessed that this would be a perfect place from which to test the SPGs and to get in some shooting practice for the ANA troops. In the afternoon the OMLT men together with a large group of ANA soldiers, headed off to Sand Hill and set up their range. Unfortunately, the SPGs could not be fired as the wrong type of ammunition had been sent with them. A useful afternoon was spent coaching the Afghans on their machine guns and personal weapons before Shaw and his men moved on to the RPGs.

Sergeant Major Mohammed, a tough and experienced Afghan soldier, was helping to coach some of the younger soldiers. This ex-Mujahedeen fighter was an excellent NCO and a strong leader. Several rockets had been fired off into the open expanse of desert and the training was going well. As each launcher fired its deadly rocket into the distance, a powerful back blast of hot gas was forced from the rear of the launchers. This together with the resultant shockwave had the effect of lifting up the top soil and creating a large dust cloud behind the firer. Everyone was startled when Mohammed suddenly screamed out, 'MINE!' The Afghan NCO had spotted a plastic bag sticking out through the disturbed top soil. Everyone knew what this meant and a short time later Owen confirmed what he believed to be a pressure plate linked to a deadly IED. It had been a miraculous escape and a chain of good fortune had saved many lives. Had the ammunition for the SPGs been correct, the weapons would have been fired from an area near the device and the crews would have been standing right on top of the pressure pad. Had the back blast from the RPGs not uncovered the plastic, who knows who could have been hurt. The device had obviously been there for some time and how no one had detonated it during Operation Chakush was a mystery. This area had been used by ISAF troops many times before.

Shaw ordered the hill to be evacuated and before long the specialists were on top of the hill cautiously feeling their way around the device. Shaw positioned his two WMIKs and an ANA ranger in overwatch positions overlooking the Green Zone, which was about 300 metres away. The clearance operation was going well, the sun was setting and Shaw stood talking to some of the EOD team just behind his WMIK. Suddenly the zinging sound of bullets filled the air and dust was kicked up all around the little party and their vehicles. Shaw could not believe the intensity of the fire; the air seemed to be thick with bullets and by the sounds they were making he knew they were passing closer than normal. The party took cover behind the exposed WMIKs and Shaw tried to move down the side of his vehicle to his radio in order to send a contact report but the sheer quantity of bullets forced him back. He knew it would be suicide to continue and suspected that bullets were passing within inches of him.

Lance Sergeant Matt Robinson and Lance Corporal Swift returned fire into the Green Zone from the GPMG and GMG on Shaw's vehicle while Lance Sergeant Green and Guardsman Hurst did the same from the second WMIK. All the OMLT men showed great courage by standing up and returning fire against such a terrible barrage of incoming rounds. Fortunately, the EOD team had been able to crawl into relative cover behind the WMIKs. Several minutes passed but the rate of incoming fire did not slack and the whole party was pinned down and in grave danger.

Shaw now managed to speak to Captain Rupert King-Evans in FOB Arnhem who told him that a couple of Viking armoured vehicles would be there shortly to assist. He asked for the location of the enemy firing points so that mortar fire could be brought to bear. The trouble was that none of those on the hill was able to identify a single enemy position. The Taliban were well concealed deep inside the vegetation in the low ground and all the troops could do was fire at likely areas. Mortars were not an option.

Just to the right of Shaw's position, a lone Afghan soldier stood in the back of his Ford Ranger and blazed away into the Green Zone with his big DShK machine gun. He was a brave man and the trapped onlookers hoped that some of the 12.7mm rounds he was firing would find their target. Owen was firing the 51mm mortar in the hope that its bombs might encourage their attackers to withdraw. Shaw knew that their luck couldn't last and they would have to move if support didn't arrive quickly. The problem was that the ATO party were all taking cover behind the WMIK and there was no way it could be moved without exposing them. Hundreds of incoming bullets continued to zing past and thud into anything that got in their way. The noise was incredible and the fear on the faces of all those sheltering behind the Land Rovers was plain to see. Fortunately the enemy fire suddenly reduced and the OMLT men used the opportunity to move Shaw's vehicle to a more covered location at the side of the hill. Right on cue the two Vikings arrived from Arnhem and provided cover for Owen's vehicle to move into the protection forded by the hillside. The tracked vehicles fired their smoke dischargers and a wall of white smoke masked the escape of the grateful little party who sped off to the safety of FOB Arnhem.

When they reached the battered outpost, the whole of 3 Company could be seen looking over the walls. They had watched nervously as the battle took place, powerless to help. The OMLT men were relieved to be back inside the relative security of the

FOB and to be able to exchange banter with their Grenadier mates from 3 Company. Everyone was still excited and the nervous energy expended in the firefight was now starting to take its toll.

As the soldiers discussed their experiences, no one was able to understand how they or the lone Afghan soldier had avoided being hit. They had all been under fire before but had never been pinned down in such precarious circumstances. As the adrenaline-fuelled buzz wore off, the troops found themselves feeling utterly exhausted. Some smoked as they mulled over the day's events. Everyone gulped down water to slake their intense thirst. Some found that their hands now shook. These were the inevitable after-effects of fear and nervous activity. The OMLT men would have liked nothing better than to crawl into their sleeping bags and fall into a deep sleep, but this was not yet possible. The expended ammunition had to be replaced and, as the sun went down, they once again clambered aboard their battered WMIKs for the short journey back to Zulu Crossing. When they arrived back at their little compound there was still no rest to be had. Weapons had to be cleaned, vehicles serviced and then a long night was spent taking turns on watch to ensure the security of the crossing.

On 23 August, Mace and Robison visited the OMLT positions and, to Shaw's great delight, they brought with them a .50 calibre machine gun. It was like bringing a gift to a birthday party; the OMLT men were overjoyed and rapidly set up the weapon where it could do the most damage to any Taliban attackers. That afternoon it was in use as the enemy launched their first all-out assault on the crossing. As the Grenadiers went about their daily routine, bullets suddenly cracked overhead and streams of bright tracers zipped over the compound roof. Loud thuds were heard as the deadly bullets slammed into the mud walls. On the roof-top the OMLT men returned the fire from their flimsy sandbag emplacements. The attack eventually petered out and the interpreters, who were monitoring local intelligence, suggested that the Taliban had sustained two casualties. This at least was good news.

It was necessary to reinforce the rooftop sangars as they were woefully inadequate. It looked as though the Taliban were not going to give up their efforts to dislodge the ANA and Grenadier troops from Zulu Crossing. The sangars could only be built up under the cover of darkness as it was too dangerous to walk around the rooftops in daylight. Most of the next evening was spent filling and transporting the heavy sandbags up onto the roof. This was exhausting but necessary work. It was as well that the sangars were reinforced because the following evening the Taliban launched a heavy and sustained attack on the crossing. At 1800 hours a tremendous rate of machine gun fire raked the compounds. This was joined by an alarming number of RPG rockets which exploded with bright flashes and loud bangs in the failing light. The Grenadiers and their ANA charges gave as good as they got and blasted away at the sources of the incoming fire. Enemy rounds thudded into the newly laid sandbags and cut across the heads of the Grenadiers who kept low on the rooftops. Even in this confused environment it was clear that the enemy were manoeuvring themselves ever closer to the crossing. Shaw called in supporting mortar fire and before long the reassuring crump of exploding 81mm mortar bombs was heard. This was joined by the louder explosions of the 105mm shells which had been brought in by the FST in FOB Arnhem. The shells could be heard whistling overhead and the buildings vibrated with each explosion as the lethal projectiles hit the ground. The mortars and artillery were adjusted onto the enemy positions and the explosions were now as close as 200 metres away for the mortars and 400 metres for the artillery. Lance Sergeant Tom Loder, who was the Javelin specialist, assisted by Hurst, was able to identify heat sources through the Javelin thermal sights. He could clearly see the enemy moving closer and, having acquired his target, he launched one of the deadly missiles which detonated on the target in a huge explosion of sparks. The pair repeated the process as other targets were identified. One group got dangerously close before the Javelin missile destroyed them. Owen raced about the rooftop position resupplying ammunition which was being fired at an alarming rate. Everyone, fearing that they might be overrun, helped by carrying ammunition around the compound and changing the white-hot barrels on the GPMGs so that they did not overheat.

The enemy attack eventually petered out and the Grenadiers again worked until midnight filling sandbags and improving their rooftop defences. All seven of the team were completely exhausted by the constant attacks and the sheer pressure and anxiety that their isolation was causing. They were all aware that fortune had smiled upon them by allowing them to escape injury thus far. No one believed that the situation could continue without casualties. First light brought no reassurance when they saw that the tarpaulin over their sleeping quarters had been penetrated by shrapnel from an exploding RPG rocket. No relief was expected for another three weeks yet and they all knew that their extraction would mark the end of the tour. They were deeply envious of some other elements of the battalion who were sitting in the safety of Camp Shorabak. The odds didn't look good, but there was no point in dwelling on it. All they could do was to remain alert, fight hard and see out the remaining weeks.

The routine of intermittent mortar attacks continued and, although they were used to the sound of the explosions, the troops retained their anxious feelings of isolation. They stood to in anticipation of attack each evening as the light was starting to fail. This was the Taliban's favoured hour because they could move into position while the target was still visible and could escape under the cover of darkness. On 27 August, as the troops lay in their rooftop positions, one of the interpreters reported Taliban activity.

Shaw and the team were sure that the enemy had occupied some concealed positions from which they had attacked previously. The young officer ordered Swift to engage the positions with the GMG. The NCO aimed and fired a series of 40mm high explosive grenades towards the target. By the increase in enemy movement it was clear that the right target had been engaged. The Grenadiers now opened up with all of their available weapons systems and Robinson directed mortars onto the area. Soon, the sound of rapidly exploding mortar rounds was heard from the target area. Mohammed looked delighted and the interpreter explained that the Taliban appeared to have taken significant casualties. No fewer than 62 mortar rounds were fired and their effect certainly looked to be devastating. The following morning, local farmers told of panicked Taliban fighters moving through their villages carrying many dead and wounded. Six bodies had been left behind which the locals had buried themselves. This was a rare piece of good news and everyone on Zulu Crossing was delighted.

Unfortunately, the enemy losses failed to deter the Taliban and on 31 August no fewer than 14 mortar rounds landed around FOB Arnhem. This was followed by a small arms attack on Zulu Crossing which grew into a full-scale battle, with the rooftop sangars once again attracting much attention from enemy snipers. The fight lasted for three hours, involved mortar and artillery fire once again and was brought to a conclusion by the US Air Force. An F15 fighter was called in, which strafed the tree line from which the enemy were firing, to great effect. The more persistent Taliban fighters had occupied a compound and, on direction from the ground, the US fighter dropped a 500lb bomb into the compound, killing everything within. It was a satisfying conclusion for the Grenadiers who watched the smoke and debris rise from the flattened target. The Taliban were certainly taking a lot of casualties.

*

August passed into September and the end of the tour still seemed as far away as ever for the isolated troops in FOB Arnhem and Zulu Crossing. On 2 September the brigade commander was due to visit FOB Arnhem and Shaw arrived at the little base during the early morning so that he might brief the commander. The Taliban decided to add to the occasion by firing rockets at the FOB. In this instance a 107mm missile exploded inside the FOB less than 25 metres from the briefing area. Lieutenant Charlie Malet, a Cold-stream Guards officer who had arrived only 24 hours earlier, was hit and wounded. Stevens, who had been standing next to him, emerged unscathed although the shrapnel embedded in the bastion wall next to his head gave him pause for thought.

The following day saw the arrival of Right Flank Company of the 1st Battalion Scots Guards to relieve 3 Company at FOB Arnhem. They were an armoured infantry unit equipped with the very welcome Warrior armoured fighting vehicles. These armoured monsters were a sight for sore eyes and a real boost to morale. The Taliban had noticed the 25-ton vehicles too and welcomed them with a barrage of mortar fire. Over the following days the Scots Guards provided two snipers to Shaw's OMLT team who quickly made their mark. The ANA soldiers were very impressed with the Warriors and delighted that the increased fire-power was on their side.

On 6 September, the biggest contact yet was played out. At about 1130 hours, just as the sun was reaching its highest point and the heat was becoming unbearable, a few cracks of gunfire were heard over the compound. This quickly developed into a huge volume of automatic fire which raked the rooftops followed by volleys of RPGs which exploded throwing clouds of dust and shrapnel into the air. Green was narrowly missed by an RPG as he moved one of the WMIKs into position. From the rooftop sangars it was clear that the heavy machine gun fire was coming from a tree line some 600 metres to the south of Zulu Crossing. The

Taliban now launched a determined assault from the west. Loder, using the Javelin thermal imaging sight, picked up no fewer than five separate groups of five men all manoeuvring towards the crossing. The machine guns in the tree line were providing fire support to the assault. The troops in FOB Arnhem were quick to rush to the aid of those at the crossing and Warriors were deployed to Sand Hill in support. Every available weapon system was fired at the attacking enemy and Loder again fired missile after missile into the manoeuvring Taliban who got dangerously close to Compound 3. Loder was using the Javelin at its closest possible range and the impact of the powerful anti-tank missiles shook the buildings. Fast jets now arrived on station and were quickly directed onto the target. They strafed tree lines only 400 metres away. Two 500lb bombs were dropped, again to devastating effect, and the Taliban, realising they were outgunned, once again faded away into the countryside.

The battle damage assessment for this contact was that the Warrior had killed 16 enemy fighters and the Javelin team had killed 12. Loder was credited with breaking up the enemy assault with his accurate and timely fire. Everyone hoped that the pasting the Taliban had received would discourage them from attacking for a while, but they were to be disappointed. The same pattern emerged the very next day with the enemy engaging the OMLT troops during the late afternoon. Loder, who had by now become an expert at identifying the Taliban in among the vegetation, quickly dealt with a group of six fighters by means of yet another deadly Javelin missile. Intelligence identified the presence of a senior Taliban commander. It seemed that the enemy were now becoming ever more obsessed with destroying the crossing. Mortars and artillery once again saturated the area and the deadly fast jets were on station quickly to carry out more strafing runs. As the aircraft lined up for the next run in, the panicked Taliban fighters ran for their lives.

Over this period the ANA had acquitted themselves well and had not caused too many problems for their mentors. There was an unfortunate incident when the Afghans again decided to fire their mortar without permission and narrowly missed a British patrol, Shaw was very angry with the Afghan commander, but fortunately no lasting damage was done. The poor standard of education among the Afghan soldiers was brought home to Shaw during a conversation with one of the ANA NCOs. The older man was amazed when, during a general conversation, Shaw explained to him that the earth was in fact round and not flat. The Afghan had genuinely believed that the earth was flat. Many of the ANA soldiers had had no education and could not read or write. The Afghan NCO's life had been one of hardship and war for as long as he could remember. He was brave and a good fighter as were many of his men and in Afghanistan; for now, that was enough.

After a brief lull in contact, the Royal Gurkha Rifles (RGR) battlegroup arrived in the area on 16 September for another clearance operation. The next day, to the elation of everyone at Zulu, the replacement OMLT Team from 2nd Battalion Yorkshire Regiment arrived. It was strange to look at the fresh faces and clean uniforms of the new arrivals but they were very welcome. Both teams set about the details of the handover in the following days, but the Grenadiers were not out of the woods yet. The operation that the RGR had commenced had begun and the OMLT/ANA were tasked to carry out a diversionary attack onto 'Objective Forecourt'. While this was taking place, the BRF would seize another crossing point about 400 metres to the north of Zulu and then two companies from the RGR were to cross over and assault their objective. The diversionary attack went well, the ANA did exactly as instructed and the RGR seized their objectives.

The operation continued apace and Shaw's men stayed at Zulu Crossing while the RGR continued their clearance. On the morning of the 20th, the Grenadiers were once again woken by the sound of heavy fire spraying over the rooftops. They had been in enough firefights by now to quickly realise that the incoming fire was British. Shaw ordered the ANA to cease fire and quickly liaised with the RGR HQ who ordered their troops on the ground to do likewise. In a very unfortunate set of occurrences, a young Afghan soldier who had been standing on the roof of his compound had been shot by a stray bullet from an action taking place some distance off. The remaining ANA men, believing the shot had come from the area of Forecourt, had returned fire. Their fire had in turn come close to a company of Gurkhas, who, recognising the AK-47 fire, had shot back at the ANA. There were some irate commanders and some harsh words but things could have been much worse. The ANA soldier had been shot in the buttocks and was in a good deal of pain; he was subsequently evacuated.

Shaw and his small team of exhausted Grenadiers finally flew out of FOB Arnhem in an RAF Chinook on 20 September. They had handed over to the Yorkshires who now had a good feel for the ground and would continue the defence of Zulu Crossing. As the lumbering chopper lifted away from the FOB, the tired team looked down at the valley below. This seven man group had survived 49 days of almost constant attacks at Zulu Crossing. They had fired more than 33,000 rounds between them and had inflicted some serious casualties on the enemy. They were on the way to safety at last and the prospect of a plane back to the UK in a few days' time. As Shaw looked around the drawn and bearded faces of his men, he was filled with immense pride for what they had done. Some of them grinned back, safe in the knowledge that a fresh meal and a shower awaited them.

17

END OF THE TUNNEL

Back at Camp Shorabak and in FOB Tombstone, the routine had continued throughout the stiflingly hot summer. OMLT teams came and went as the companies deployed on operations and changed their locations. Small groups often passed through and used the opportunity to fill their vehicles with water, ammunition and every little luxury that they could lay their hands on. Cans of fizzy pop, sweets and chocolate were always very welcome after the austerity of the patrol bases. Anyone staying overnight got the added bonus of fresh food and a shower. These visits were also used to collect as much mail as possible. Parcels and letters were regularly despatched by helicopter to the FOBs, but unfortunately mail was a much lower priority than some of the other stores and equipment. The mail sacks often sat near the helipad in Bastion for long periods of time, much to the frustration of the men in the FOBs. While the OMLT teams in the field fought alongside their Afghan colleagues, the troops in Shorabak and its little annex in FOB Tombstone went about the business of sustaining both the British OMLT men and Afghan soldiers in Helmand. Training and mentoring went on daily and the continual process of improving the brigade HQ staff and logistics personnel remained challenging. Plans were now moving on and a fourth infantry kandak was being raised and trained. There were no extra mentors for this task and all available personnel were roped in to assist.

Visits to Tombstone were few and far between for most of the troops mentoring kandaks and the commanding officer was aware that most of the men had seen some heavy fighting during their time away. Prolonged exposure to the stresses of combat can have profound effects on some soldiers and Lieutenant Colonel Carew Hatherley was determined to ensure that the psychological welfare of his troops was taken care of. An effective psychological risk assessment programme known as trauma risk management or TRiM, had been adopted by the Grenadiers some years earlier.* A large number of officers and NCOs had been trained in how to identify the signs and symptoms of post-traumatic stress and the system was now working well in Afghanistan. Soldiers who had been through difficult situations were routinely interviewed by their own NCOs and an assessment was made of how they were coping. Their progress was then monitored and any deterioration in their wellbeing could be quickly identified and the men concerned referred to medical help. Few men needed qualified medical intervention, but the system was in place to catch those who might find themselves in difficulty subsequently. Most found the process helpful, although there was nothing that could blot out some of the more distressing incidents from the mind. Many of these assessments were carried out in FOB Tombstone during the brief periods of respite from the desert environment.

Any soldier from the Grenadiers or the OMLT teams admitted to the field hospital in Camp Bastion was visited by the commanding officer and other personnel from Tombstone. The padre was a frequent visitor as was the Grenadier's own medical officer. Serious casualties were always moved back to the UK at the earliest opportunity but there were plenty of sick men as well as wounded

---

* Originally a Royal Marines initiative, the Grenadier Guards pioneered the use of the system at unit level in the Army. TRiM was later formally adopted by the Army and a Grenadier officer was placed in charge.

to be looked after. The unenviable task of identifying the dead from the OMLT and Grenadier sub-units usually fell to personnel from HQ and was a gruesome experience that would live long in the memories of those who carried out this duty.

The repatriation ceremonies had become all too familiar; by the end of August, 22 British servicemen had been killed in southern Afghanistan since the first fatality on 13 April. Most of these young men were repatriated from Camp Bastion. The entire camp population paraded on the dusty airfield to watch as the coffins draped in union flags were solemnly marched up the ramp of the waiting RAF aeroplanes. These were moving and necessary ceremonies, but they were all too frequent and there would sadly be many more before 12 Brigade returned to the UK.

There were many comings and goings to the little camp in FOB Tombstone. Battle casualty replacements arrived and were despatched to their companies along with other arrivals. Various appointments had changed and among these was Major James Bowder who took over command of 2 Company mid-tour. Soldiers departing and returning from R&R also transited through the little Grenadier base. There were many visits from VIPs and senior officers during the summer and even the British ambassador to Afghanistan found the time to call in at Tombstone, which he declared was the cleanest camp in the country.

While 2 Company had been fighting around Zulu Crossing in late July, the Queen's Company had been recovering in FOB Tombstone and had been helping to regenerate the battered 1st Kandak. It had been a much needed break and some members of the company were just returning from leave in the last days of July. Major Martin David returned to Afghanistan on 3 August and only a day later found himself deploying with his company on yet another challenging mission. 3rd Kandak and the Inkerman Company had been in Sangin for an extended period and were now due relief. The Queen's Company was to replace them in the notoriously dangerous town. Just as 2 Company men were taking over the defence of Zulu Crossing, their colleagues in Tombstone prepared to relieve the Ribs in Sangin. Unusually, the Queen's Company and most of their Afghan troops were flown into Sangin on four Chinooks all at the same time. Everyone knew that the British force was short of these precious helicopters and to see four all at once was a boost. By now the men were more than used to arriving in new hostile environments, although Sangin had a fearsome reputation and there had been many casualties there over the preceding months. The Queen's Company men had seen their fair share of action and were under no illusion as to the danger they would be in for this latest deployment, especially as it would probably see them through to the end of the tour. The company had been reorganised due to changes in personnel and casualties; this meant they would not all be working with the Afghans they had become familiar with in recent months, which was frustrating for some.

Over several days, the men of the Inkerman Company handed over to their colleagues in the Queen's Company who familiarised themselves with the Sangin area of operations. The distribution of the troops had not changed markedly and the OMLT still manned the Sangin PBs – Tangier, Waterloo and Blenheim as well as a small detachment in Inkerman. There was roughly a platoon of Afghans at each location. A Company of the Royal Anglians was now based in the Sangin district centre and David was able to make some improvements to the command and control of the OMLT team there. Closer links were forged with the Anglians and an OMLT operations room was established which tapped into the improved communications that the infantry company had from the district centre. Sustaining the Afghans and their mentors in the isolated PBs remained a real problem with the scant resources available. Stores, rations and equipment of all sorts were flown into Sangin district centre almost daily by the

Chinooks. Once unloaded, the equipment had to be sorted and broken down to ensure that it got to the right places. Resupply of the PBs was still conducted by dangerous patrols through the town in the battered WMIKs and even Snatch Land Rovers, which had been designed for use in Northern Ireland. Water, food, batteries, mail and ammunition all had to be delivered in this way and if a patrol was cancelled then the PBs would have to ration themselves until the resupply run got through. It was a ridiculous situation given that the distance between the locations in Sangin was just a few kilometres at most, but it demonstrated the severity of the threat against any patrols moving out of the FOBs. The logistic situation was complicated by the needs of the Afghans who insisted on eating fresh meat and cooking over open fires. Goats and firewood were sometimes transported to the little outposts just to maintain Afghan morale. About every six weeks a major logistic convoy was put together. All the heavy stores and building materials required to sustain the deployed force would be driven to Sangin via FOB Robinson. Shortly after the arrival of the Queen's Company, two months' supply of wood for the Afghan camp fires was driven up, which was a great deal of timber. These logistic patrols were major undertakings, not least because convoys were usually hit by IEDs on the way out and on the return journey too.

To the north of Sangin, the isolated little outpost known as PB Inkerman had been reinforced and in addition to housing the small ANA detachment it was now the temporary home of C Company of the Royal Anglians. The increase in manpower had done little to improve the situation there and the small compounds were under attack generally twice a day. The enemy were targeting the area used as a helicopter landing site and ISAF had restricted flights because of the threat. As if this were not bad enough, the Taliban had continued to lay IEDs and booby traps along Route 611, further restricting movement and adding to the isolation of the little base. Having pretty well fixed the British force in place, the Taliban now increased their attacks on the FOB to try and squeeze the British out. Any patrol leaving the base usually found itself under attack within minutes. On the day before the Queen's Company arrived to relieve the Ribs, the Royal Anglians lost another man to enemy action not far from the base.

Paddy Hennessey, recently promoted to Captain, was placed in charge of the Amber 63 group which moved up to FOB Inkerman. His OMLT had been drastically reorganised. Some of his most experienced men, who had seen a great deal of action, were cut across to the other platoons. The Queen's Company men set about installing themselves with their Afghan counterparts from 1st Kandak. Unfortunately, the ANA had occupied the area at the bottom of the slope, near the road. This meant that any mortar or rocket fire that was directed into the base from the Green Zone went over the top of the ANA buildings. Any mortar bombs falling short invariably landed on or close to the ANA and OMLT positions. To visit the operations room it was necessary to move across an open area uphill. This was a dangerous occupation given the frequency with which RPGs, rockets and small arms fire landed in the area. The journey was soon referred to jokingly as the 'Dash of Death'. The shrapnel marks and bullet holes in the walls bore testament to the amount of ammunition that fell on the OMLT buildings. The large hole made by the SPG 9 that had come so close to killing the Inkerman Company men in their temporary accommodation weeks earlier was still there and didn't do anything to improve the confidence of the new occupants.

The first few days in FOB Inkerman were punctuated by frequent rocket and small arms attacks. Like the Inkerman Company before them, the Queen's Company fought from the sandbagged rooftops using all of their weapons systems. RSM Andrew Keeley put in another appearance and was seen firing magazine after magazine from the roof alongside the men. The added firepower provided by C Company was very welcome and the Grenadiers and Royal Anglians fought alongside each other once again. The Taliban crept in as close as they could to fire their RPGs and machine guns, but it was the lethal 107mm rockets that caused the most damage. They were not particularly accurate, but when they landed their deadly shrapnel showered the inside of the FOB. On 11 August the Taliban mounted a particularly fierce set of attacks on FOB Inkerman. Everyone available fought from the roof tops and returned fire at their attackers. RPGs, mortars and rockets were all used and the British blazed away in response with everything they had. It was a fierce fight and another British officer was killed.

A week later the Taliban again launched a prolonged and determined attack on the FOB. Rockets and mortars rained down at an increasing rate. The missiles made an ear-shattering noise when they exploded nearby, causing everyone to duck . At one point Hennessey, accompanied by Guardsman Gillespie and a couple of others, dashed across the open interior of the FOB as a 107mm rocket impacted far too close to them. Those who were not blown off their feet dived for cover and thanked their lucky stars that they had not been vaporised. Everyone's ears rang and a cloud of dust hung in the air. The rocket casing could be seen nearby and the battle continued to rage around the little party. Hennessey ran on and only realised when informed by the Afghans that Gillespie had been hit. When the Grenadier officer returned, Gillespie was being treated. He had been blown sideways and had taken a lot of shrapnel. There were extensive injuries to his arm and face but the young soldier was conscious and talking, which was a good sign. Before long he was bandaged on a stretcher and was smoking a scrounged cigarette. It seemed like an age before the MERT helicopter arrived, but it looked as though Gillespie would make it, even though his facial injuries appeared to be severe.

Back at Sangin district centre, David was directing OMLT operations and ensured that each of the three PBs mounted two or three patrols during every 24-hour period, even during Ramadan, which was not popular with the ANA. These patrols were frequently contacted by the Taliban who appeared to be determined to maintain the pressure on Sangin. Company HQ mounted occasional patrols in the bazaar area but a lack of transport restricted movement much further from the district centre. The troops in PB Tangier frequently came under fire from an area known simply as 'Tank Park'. This was a collection of rusting Soviet armoured vehicles left behind by the defeated Russians. The contacts normally ended badly for the Taliban as the OMLT response was generally aggressive and effective. Jets were often used to great effect, and the Taliban were terrified of the fast-moving aircraft and their deadly bombs; they knew there was nowhere to hide once their locations had been fixed. However, the perils of cooperating with aircraft when in close contact with the enemy were brought home on 23 August. Near Kajaki, a patrol from the Royal Anglians had requested air support and two American F-15 fighters dropped bombs onto the Taliban positions. Tragically, one of these bombs fell onto a compound occupied by British troops and three young soldiers were killed.

The Amber 61 troops in PB Blenheim also hurt the Taliban badly on a number of occasions. The Javelin thermal imaging control unit again proved its effectiveness at night when a Taliban patrol of seven men was seen approaching to attack. A missile was launched and all the enemy fighters were killed. The men in PB Waterloo were still responsible for securing the southern entry route into Sangin. IEDs were laid almost every day as the Taliban attempted to deny the route and the Grenadiers were grateful that the ANA were so efficient at spotting these lethal devices. During the final eight weeks in Sangin the OMLT and ANA dealt with around 35 IEDs in the Sangin area alone. Everyone became very proficient in spotting the 'bomb layers' and some effective ambushes were laid. This pattern of daily Taliban attacks and ISAF counter-strikes continued throughout August and into September with neither side showing any obvious signs of giving up the fight.

On the evening of 11 September, Amber 64 at PB Waterloo deployed a patrol which was to overwatch the southern approaches to the town. They had previously had some success in ambushing the layers of the IEDs and this activity was an important part of defending the routes. As the team patrolled towards their objective, there was a sudden huge explosion and a blinding flash. An ANA sergeant had triggered a large mine which blew him off the ground causing a traumatic amputation of his legs. The interpreter too was hit in the head and seriously injured. Lance Sergeant 'Goolie' Ball moved to take up a fire position and tragically triggered a second concealed device. Ball was thrown to the ground having lost most of his left leg. Second Lieutenant Tom Hamilton found himself dealing with three serious casualties on one of his first patrols in Afghanistan. To make matters worse, the young officer had also taken shrapnel to his own thighs, buttocks and back. Rapid first aid was given to all the casualties. Guardsman Arnie Snyman, another new arrival, applied two tourniquets to what remained of Ball's shattered leg. This was no mean feat in the dark with the threat of further IEDs in the immediate area. It was important now to get the casualties extracted quickly. They were loaded onto vehicles which raced off towards the district centre where every available medic had been crashed out to receive them. As the casualties were receiving first aid, the wounded interpreter slipped into unconsciousness and died. The wounded ANA man and Ball were stabilised and the worst of the bleeding was stopped. David held Ball's hand and tried to reassure him, but it was clear the wounded NCO was in pain and more morphine was administered. It took around

40 minutes for the MERT chopper to arrive and the Queen's Company men were greatly relieved when it finally lifted off for Bastion with the seriously injured men aboard.

The injury to Ball, only 11 days before the Queen's Company was due to end the tour in Sangin, was a dreadful blow to morale. It seemed somehow worse than all of the other casualties the company had taken because it was so close to the end. Everyone had prayed that they might make it out without any further casualties and this had shattered any illusions of safety. More to the point, with such a short time left no one wanted to take any unnecessary risks. Some questioned the need to continue the aggressive patrol pattern when the end of tour was so close and felt that their luck was finally deserting them. David knew instinctively that strong leadership was required to maintain morale. He reminded everyone that they made their own luck and that all of their previous sacrifices would be in vain if they lost their nerve now. He pointed out that the Grenadiers' own regimental motto *'Honi soit qui mal y pense'* (Evil be to him who evil thinks) was exactly what was going to be done. The Taliban had struck an evil blow and the Grenadiers would now renew the fight with vigour until the end of the tour when they were ordered to leave their posts.* There was a further sobering incident when, three days later, a patrol from PB Waterloo discovered Ball's severed foot lying in the road. It had been wrapped in cloth and the interpreter's perception was that the foot had been placed in the road by the Taliban as a message to the British troops. It seemed wrong to simply leave a part of their friend there, so it was decided to take the foot back to the PB. Once back at Waterloo the Grenadiers solemnly cremated it; this seemed like the right thing to do, but it was a gruesome and strangely emotional event.

---

* Major David was later awarded the Military Cross for his outstanding leadership during the tour, notably the action at Adin Zai.

These final weeks took much determination and indeed inno-vation to avoid setting routines and allowing the enemy the opportunity to lay an ambush. More aggressive operations were undertaken to the north, around FOB Inkerman, and the arrival of the BRF and an Estonian group in the enemy's rear had the desired effect. C Company was reinforced by a troop of Mastiff armoured carriers which significantly improved mobility in the area. The ANA positively adored these huge-wheeled vehicles with their air-conditioning and heavy armour. Until now the only OMLT vehicle in FOB Inkerman had been a battered Snatch Land Rover with a bullet hole through the windscreen.

Operation Palk, the latest aggressive Task Force Helmand operation, was now under way. More reinforcements arrived at FOB Inkerman in the form of elements from 3rd Kandak and their Inkerman Company mentors from Shorabak. It was now possible to push into the Green Zone and move the fight to the Taliban's back yard. For the Inkerman men, this felt like going over the same ground again. One thing was for sure at this stage of the tour, no-one was taking any silly risks.

3 Company, who had been withdrawn from FOB Arnhem, discovered that there was to be no extended rest period for them. On 6 September they flew back to Garmsir, the southern town that they knew so well. They were to see out the rest of the tour there. The troops had hoped for a longer period in Camp Bastion but they knew Garmsir well and it was like coming home. FOB Delhi was still far from luxurious, but after the weeks spent in FOB Arnhem things could have been a lot worse. A Company of 2 Mercian had been in Garmsir since the Grenadiers left and they were waiting to move north again. The Mercians had decided to use this opportunity to mount a final operation designed to push back the Taliban forward positions. 3 Company's role would be a supporting one; they would split into three groups:

one to guard FOB Delhi, one to take over the checkpoints that were so familiar to them and one to act as a QRF in case things didn't go well.

A Company patrolled out from the FOB at about 2000 hours on 8 September. Those in the checkpoints and on JTAC Hill strained their eyes as they tried to monitor the Mercians' advance. It was exceptionally dark and the night vision devices were next to useless. For four hours all remained quiet and the going was incredibly slow due to the poor light levels. Suddenly, at around 0100 hours, all hell broke loose and a tremendous firefight started. Tracer rounds arced through the air and the sound of automatic gunfire echoed across the small town. Those in FOB Delhi found it difficult to follow the battle; the situation was very confused, but before long it became clear that there were friendly forces' casualties. The fighting went on for what seemed an age and the Mercians were obviously in a very difficult situation and taking a steady stream of casualties. 3 Platoon acted as a QRF and moved forward to assist with the Mercian wounded as and when they could. Sadly, there were fatalities. The others were ferried to the rear but it was obvious that extracting them in the poor light and under fire was extremely difficult. The Mercians eventually withdrew from the immediate area, but the operation to evacuate casualties went on until first light. Soldiers from several units including the Grenadiers helped as they could. The skies were full of aircraft waiting to pounce on any Taliban interference and the Light Dragoons brought up their Scimitar reconnaissance vehicles in support. The operation had gone badly and A Company had taken a lot of casualties that night; 3 Company had been in similar circumstances a few months earlier and they knew exactly what the Mercian soldiers were feeling now.

With the departure of the Mercians, 3 Company fell back into their own familiar routine in Garmsir. The checkpoints and JTAC Hill remained the most dangerous areas and the Taliban continued their attacks as before. There were daily attacks by small arms and RPGs; it was as if the Grenadiers had never been away. Worryingly, the Taliban were now able to direct mortar fire directly into FOB Delhi and the town. These attacks were not too frequent and although they were dangerous, they were nothing compared with the intensity of fire the troops had endured at FOB Arnhem. The enemy had now sited three anti-aircraft guns in the centre of their positions and the Grenadiers tried hard to locate them. The Taliban kept them well concealed and it seemed they were intent on using them to bring down a helicopter. They had no intention of giving their positions away until it suited them. It was consequently impossible to positively identify them.

On 19 September, six Afghan police casualties and three civilians including a small child were brought FOB Delhi for treatment. In the initial confusion the Grenadiers were told that a suicide bomber had detonated at a police checkpoint. The subsequent investigation revealed that an accident with an RPG had caused a number of casualties. Many of the wounded were in a serious condition and it was a grizzly morning's work for those in FOB Delhi. Shortly after, General Dan McNeil, a senior US officer and commander of ISAF, visited the base. The general, together with his entourage, arrived in Garmsir in two Blackhawk helicopters. The sight of two American helicopters immediately alerted the Taliban to the presence of a senior commander. A little later the General visited JTAC Hill and as the ground was being pointed out to him, several enemy bullets impacted in the sandbags in front of him. It was necessary for his close protection team to haul him to safety, but the Vietnam veteran was unfazed by the experience, joking about it afterwards. Such was the routine for the men of 3 Company who now discovered that they would be the last from the battalion to leave Afghanistan.

The OMLT troops, who had been among the first to deploy, were to be recovered to the UK at the end of September. In

Camp Shorabak the 2nd Battalion Yorkshire Regiment began arriving to commence the handover. The Grenadiers tailored some training to give their replacements the best possible start to the tour, but were pleased to see that the Yorkshire men seemed to be better prepared than they had been seven months earlier. There were now more specialist weapons and equipment available for training in the UK and the Yorkshires were consequently in good shape. They had a lot to learn, though, just as the Grenadiers had back in the spring. Every effort was made to provide the information and expertise that would be needed by the new arrivals. It wasn't long before replacement OMLT teams were leaving Shorabak for the outstations. Clean-shaven, keen and anxious-looking soldiers mounted the Chinooks to fly north to Sangin and the other isolated locations such as FOB Arnhem. A few days later, the same helicopters touched down in Camp Bastion and down their ramps walked a series of bearded, filthy and exhausted men, their bergans hooked over one shoulder and their dusty weapons gripped in their free hands. Despite initial appearances, they were all grinning wildly. They were safe. They had survived.

Earlier, as the Grenadiers had left their outstations they had said their goodbyes to the Afghans that they had spent more than half a year fighting alongside. It had often been a difficult relationship, frustrating and at times very worrying, but there was a mutual respect. However hard the British had fought, they could not disguise the fact that they were going home and that their Afghan colleagues were staying on for more tough fighting ahead. The attrition rate on the ANA had been very high. Scores of Afghan soldiers had died and they deserved the respect of the British. The Afghans seemed genuinely sorry to see their Grenadier mentors go and there were warm handshakes and hugs before the Grenadiers handed over to their slightly worried-looking replacements from the Yorks.

END OF THE TUNNEL

As they flew over the vast Afghan desert, each man had his own thoughts: lost friends, savage fighting, near misses and the memory of the anxiety that they might not make it through to the end. Dusty faces wore broad grins as the helicopters approached Camp Bastion. As the helicopter put down on secure sand, a great feeling of relief came over the Grenadiers as they realised they had made it.

POSTSCRIPT

The OMLT men from the 1st Battalion Grenadier Guards were all back in the UK by 26 September. The BRF and 3 Company men had rather longer to wait and didn't make it home until mid-October. For Lieutenant Colonel Carew Hatherley, the three week delay until his battalion was reunited in Aldershot was agonising. While the OMLT men were safe at home with their families, the men of 3 Company and the BRF were still very much in harm's way. To have taken further casualties at this stage would have been a dreadful blow. Mercifully, the remainder of the battalion made it home without further loss. As the coaches returning the Grenadiers to Aldershot arrived in Lille Barracks, the disembarking soldiers were reunited with their families. Weeping wives and girlfriends hugged their loved ones and soldiers choked back tears of their own as they held newborn babies and rapidly growing toddlers. Small children looked on confused as tall men in desert combat uniforms hugged their mothers. There were union flags, and sheets painted with the words 'Welcome Home Daddy', no doubt mass produced in the local junior school. This was a happy time, a time of relief; there was no place for reflection now, that would come later. Dotted around the place were a few men with walking sticks and obvious scars, casualties from the early part of the tour who were on the road to recovery and pleased to see their mates home safe.

The battalion had returned to the UK via a brief spell in Cyprus. It was mandatory for all troops to undergo a period of what the UK forces refer to as 'decompression'. The aim was to allow troops a graduated return to normality. There had been time on the beach, lectures on post-operational stress, safe driving and so on. There was also the first taste of beer for nearly seven months, all in a controlled environment. Some were sceptical about the value of this process and just wanted to get home, but most recognised that it was useful to 'chill out' a little before being reunited with families. The battalion welfare officer Captain Neil England and his team knew only too well how difficult the reintegration would be for some. Captain England had spent seven months looking after the welfare of soldiers and their families. He had repeatedly visited the bereaved and the wounded in hospital and had seen the emotional impact of war on dozens of families. The welfare team were perhaps the unsung heroes of the tour.

As the soldiers were reunited with their loved ones, the memory of those who didn't come home was never far from their minds. November 5 is the anniversary of the Battle of Inkerman, where in 1854 the 3rd Battalion Grenadier Guards fought for their lives against the advancing Russian hordes. This date is traditionally celebrated by the men of the Inkerman Company, but 5 November 2007 was especially poignant. A large group of men from the Ribs travelled to the National Memorial Arboretum where they laid a wreath to the fallen. Afterwards they moved on to Manchester and were hosted by the family of Guardsman Downes who was so tragically killed earlier that summer. There were others, of course, whose lives had been changed through injury. Some, like Lance Sergeant Ball, were still in the military wing of the Selly Oak Hospital in Birmingham and others were undergoing further treatment at the Defence Medical Rehabilitation Centre at Headley Court in Surrey. Some, like Alex Harrison who had been shot in the head in Kajaki, made incredible recoveries. The road to wellness would be a long one for most of them, but they were young, fit and incredibly determined. Some recovered sufficiently to deploy to Afghanistan again in 2009. Many of the amputees continue to serve their country in the Grenadier Guards more than four years later.

On Thursday 8 November 2007, Padre Dunwoody addressed the whole battalion and many of their families in the Garrison Church at Aldershot. It was an especially moving remembrance service and was probably the final organised event of the tour. It would be another 18 months or so before the battalion saw Afghanistan again. On 30 November the battalion dispersed on its post-operational tour leave. The events that took place in Helmand in the summer of 2007 were now mere memories, but for some they would remain painful for years to come. The men who died in Helmand would not be forgotten by the Grenadiers, but already minds were turning to the path ahead and where developments in the dusty interior of Afghanistan would lead. Most men fully expected to return to the Helmand River Valley and to trade blows with the Taliban again.

ACKNOWLEDGEMENTS

This book is first and foremost a story about the real experiences of soldiers. It is primarily about Grenadiers and it was to Grenadiers that I turned for information, but I was given many superb accounts by those who served courageously alongside the 1st Battalion Grenadier Guards. It would simply not have been possible to record these events without the patience and cooperation of so many people. Memories rapidly fade and I have repeatedly asked soldiers and veterans to dig deep into sometimes quite painful recollections. I simply cannot mention everyone who has helped with the project for they are simply too numerous. Often a small nugget of information from a Guardsman has been vital to fill in the gaps left in other pieced-together accounts. The people that gave me these vital snippets will probably never realise how significant they were but I am none the less eternally grateful to them. Most of the information has come from the soldiers of the 1st Battalion Grenadier Guards and I am grateful to Lieutenant Colonel Carew Hatherley for allowing me to commence the project and for his successors in enabling me to complete it. Captain Andrew Keeley was an invaluable ally who assembled information and arranged many interviews for me. A formidable 'fixer', his memory too proved to be quite excellent in establishing who was where at any one time.

The photographs in this book were also assembled from various sources, but my thanks must go to Marco Di Lauro and to Getty Images for allowing me to reproduce some of Marco's excellent shots. Alexander Allan, who is mentioned in the very first chapter of this book, kindly allowed me to use many of the first-class images that he took during the tour. There are many others who individually sent me photographs, maps and pictures; I could not have completed this project without their kind assistance.

The key information contained in these pages has come mainly from individual accounts which have been sent to me from many sources. I cannot mention them all but must list some of the more prominent authors of notes provided for me and some of the interviewees: Martin David MC, Patrick (Paddy) Hennessey, Rob Worthington, Howard Cordle, Rupert Stevens, Rupert King-Evans, Marcus Elliot-Square, Carew Hatherley, Andrew Keeley, Ian Farrell, Chris Gilham, Andrew Tiernan, Scott Roughley, Wayne Scully, Darren Westlake, Darren Chant (later tragically killed in action), Mark Gaunt, James Shaw, Piers Ashfield, Howard Lawn, Elliott Hennell, Rob Pointin, Tim Leatherland, Chris Bangham, Paul Fear, Simon Edgell, Jack Mizon, Ed Janvrin, Phil Hermon, Andrew Robinson, Clint Gillies, Andy Hill, Dan Malcangi, Ty-Lee Bearder, Bryan De-Vall, Dave Groom, Dave Harrison, James Shaw, Matthew Betts, Carl Shadrake, Stephen Hodgson and Milan Torbica.

My thanks go out to the scores of other contributors; to those who have continued to encourage me; and to my family for once again putting up with my mounds of paper, files, maps and general distraction.

GLOSSARY OF MILITARY TERMS AND ABBREVIATIONS

ANA	Afghan National Army
BRF	Brigade Reconnaissance Force
CQMS	Company Quartermaster Sergeant
CS	Combat support
CSM	Company Sergeant Major
CSS	Combat service support
D&V	Diarrhoea and vomiting
DShK	(*Degtyaryova-Shpagina Krupnokaliberny*) Large Soviet-era 12.7mm heavy machine gun
EOD	Explosive ordnance disposal
FOB	Forward Operating Base
FST	Fire Support Team
GMG	Grenade machine gun
GPMG	General purpose machine gun
Humvee	US Army armoured utility ehicle
IED	Improvised explosive device
ISAF	International Security Assistance Force
Javelin	Anti-tank missile
kandak	Afghan battalion (rough equivalent)
LEWT	Light Electronic Warfare Team
Mastiff	Heavily armoured wheeled personnel carrier
MERT	Medical Emergency Response Team
NATO	North Atlantic Treaty Organisation
OMLT	Operational Mentoring and Liaison Team
PB	Patrol base

Pinzgauer	Six-wheeled British Army utility vehicle
PKM	PK machine gun
QRF	Quick reaction force
REME	Royal Electrical and Mechanical Engineers
RGR	Royal Gurkha Rifles
RPG	Rocket propelled grenade
RSM	Regimental Sergeant Major
RSOI	Reception, Staging, Onward-movement & Integration
sangar	Emplacement protected by HESCO or sandbags
Snatch	Lightly armoured Land Rover
TA	Territorial Army
UAV	Unmanned aerial vehicle
VCP	Vehicle checkpoint
Viking	Tracked armoured personnel carrier
wadi	Dry river bed
WFR	Worcester and Sherwood Foresters Regiment
WMIK	Weapons Mount Installation Kit (on a Land Rover chassis)
Zulu muster	Holding area for vehicles